Psychology zone

shortcuts to be[...]

AQA
YEAR 2 A-LEVEL
Psychology

BRILLIANT MODEL ANSWERS

Forensic Psychology

- Provides the key knowledge and skills for exam success
- All types of questions covered
- Grade A/A* model answers
- Written by examiners

Do brilliantly in your Psychology exam!

Nicholas Alexandros Savva

psychologyzone.co.uk

| Proven exam success | Written by examiners | Concise, detailed and clearly written model answers |

Brilliant Model Answers

Published by

Educationzone Ltd

London N21 3YA
United Kingdom

©2021 Educationzone Ltd

All rights reserved. The copyright of all materials in this publication, except where otherwise stated, remains the property of the publisher and the author. No part of this publication may be reproduced, stored in a retrieval system or transmitted, in any form or by any means, for whatever purpose, without the written permission of Educationzone Ltd or under licence from the Copyright Licensing Agency, the 5th Floor, Shackleton House 4 Battle Bridge Lane London SE1 2HX.

Nicholas Savva has asserted his moral rights to be identified as the author of this work in accordance with the

Copyright, Designs and Patents Act 1988.

Any person who commits any unauthorised act in relation to this publication may be liable for criminal prosecution and civil claims for damages.

British Library Cataloguing in Publication Data:

A catalogue record for this publication is available from the British Library.

978-1-906468-10-1

Email us for further information:

info@psychologyzone.co.uk

For more information:
Visit our website for exam questions and answers, teaching resources, books and much more:
www.psychologyzone.co.uk

Content for Forensic Psychology

Important information 3

Exam skills 4

Specification: Forensic psychology 7

Offender Profiling: Top-down approach 8

Offender Profiling: Bottom-up approach 12

Biological explanations of offending behaviour: a historical approach 16

Biological explanations of offending behaviour: genetics and neural explanations 20

Psychological explanations of offending behaviour: Eysenck's theory 27

Psychological explanations of offending behaviour: Cognitive explanations 31

Psychological explanations of offending behaviour: Differential association theory... 36

Psychological explanations: Psychodynamic theory 40

Dealing with offending behaviour: custodial sentencing 44

Dealing with offending behaviour: behaviour modification 50

Dealing with offending behaviour: Anger management 54

Dealing with offending behaviour: Restorative justice 58

Answers to identification questions 62

Please note: this book is not endorsed by or affiliated to the AQA exam board.

Important information

⚠ Do not skip this page!

■ Isn't the exam supposed to be unpredictable?

This guide is part of Psychologyzone's *Brilliant Model Answers* series covering A-level Psychology. Use it alongside the Psychologyzone series Brilliant Exam Notes to get the best out of your learning.

This guide to the 'Forensic Psychology' topic provides a full set of exam-style questions and model answers to help you do well in the exam. After all, your Psychology exam is based on answering questions – what better than to have a book that already has the answers for you?

The exam board has deliberately developed the A-level Psychology specification so that the questions are to some extent 'unpredictable' in order to discourage students from attempting to rote-learn (memorise answers) using pre-prepared questions. This makes it difficult to predict what's going to be asked.

We have tried to make the unpredictable 'predictable'.

There are over 90 model answers in this book. We have covered most of the different types of question they can ask you for each topic on the specification. You can adapt the model answers provided to most types of questions set in the exam.

■ Some of your model answers seem very long. Why?

Some of the answers are much longer responses than you would need to write in the exam to get top marks. **This is deliberate.** We have written them this way to enable you to have a better understanding of the theories, concepts, studies, and so on. If you do not write as much as we have, don't panic! You don't need all of the content to achieve a good grade.

As you may be using this as a study book, we thought we'd write the model answers in a way that means you can also revise from them, so we sometimes expand on explanations or give an example to help you understand a topic better.

Many of the model answers start by repeating the question; in the real exam you don't need to waste time doing this – just get stuck in!

> Remember: in your exam, your answers will be marked according to how well you demonstrate the set assessment objectives (AOs). We have tried to provide model responses that show you how to meet these AOs. Each example provides you with 'indicative content' – in other words, the response gives you an idea of points you could make to achieve maximum marks. It doesn't mean these are points you must make! The purpose of these model answers is to inspire you and demonstrate the standard required to achieve top marks.

Exam skills

How will my answers be assessed?

Your teachers will have explained that your answers in the examination will be assessed on what examiners call assessment objectives (AO). If you can familiarise yourself with these AOs, this will help you write more effective answers and achieve a higher grade in your exam. There are three assessment objectives: AO1, AO2 and AO3.

By now, your teachers should have given you a lot of practice exam questions and techniques for how to answer them. The aim of this book is not to teach you these skills, but to show you how it's done – to model the answers for you.

Just to remind you, below are the AQA assessment objectives:

AO1 Knowledge and understanding

Demonstrate knowledge and understanding of scientific ideas, processes, techniques and procedures.

What does this mean?

The ability to describe psychological theories, concepts, research studies (e.g., aim, procedures, findings and conclusions) and key terms. The exam questions can cover anything that is named on the specification.

Example

Explain the process of synaptic transmission. [5 marks]

Outline the role of the somatosensory centre in the brain. [3 marks]

AO2 Application

Apply knowledge and understanding of scientific ideas, processes, techniques and procedures:
- in a theoretical context
- in a practical context
- when handling qualitative data
- when handling quantitative data.

What does this mean?

Application questions require you to apply what you have learnt about in Psychology (theories, concepts and studies) to a scenario (situation) often referred to as 'stem' material. A scenario will be a text extract or quote given in the question. You are treated as a psychologist, and you need to explain what is going on in the situation from what you have learnt.

Example

Chris suffered a stroke to the left hemisphere of his brain, damaging Broca's area and the motor cortex. Using your knowledge of the functions of Broca's area and the motor cortex, describe the problems that Chris is likely to experience. **[4 marks]**

  Evaluation

Analyse, interpret and evaluate scientific information, ideas and evidence, including in relation to issues, to:

- make judgements and reach conclusions
- develop and refine practical design and procedures.

What does this mean?

Evaluation simply means assessing the 'value' (hence 'evaluation') of a theory or study you have been describing. There are many ways you can evaluate theories or studies. For students, evaluation often takes the form of the strengths and weaknesses of the theory and/or study, but evaluation can also be in a form of 'commentary' (neither strength nor weakness but more in the form of an 'analysis', which is still an evaluation).

Example

Outline one strength and one limitation of post-mortem examination. **[2 marks + 2 marks]**

What are the different types of exam questions?

We have grouped the exam questions into four different types:

Identification questions	Multiple-choice questions, match key words with a definition, tick boxes, or place information in some order or in a box.
Short-response questions	Questions worth up to 6 marks (1, 2, 3, 4, 5 or 6 marks). These are often questions asking you to 'outline', 'explain', or 'evaluate' a theory or a study.
Application questions	These require you to apply the psychological knowledge you have learnt (theories, concepts, and studies) to a real-life scenario given in the exam question.
Long-response question	These questions require longer answers and are worth over 6 marks (8, 12 or 16 marks). The long-response answers found in this book will be mainly for 16 mark questions.

How are the model answers structured?

We have tried to structure your learning by breaking down the model answers into four distinct categories:

Key terms, concepts, and **theories** that are named on the AQA specification are covered by the identification and short-response questions (e.g. explain what is meant by the term...).

Research questions asking you to outline a study, describe a theory or give an evaluation are covered by short-response questions (e.g. briefly outline one study that has...).

Application questions require you to apply your knowledge to a made-up scenario (situation) and are covered under application questions.

Essay questions 'Outline and evaluate', or 'Discuss'-type questions are covered under long-response questions. Some long-response questions also require the application of knowledge.

Specification: Forensic psychology

Forensic psychology

- Offender profiling: the top-down approach, including organised and disorganised types of offender; the bottom-up approach, including investigative Psychology; geographical profiling.

- Biological explanations of offending behaviour: an historical approach (atavistic form); genetics and neural explanations.

- Psychological explanations of offending behaviour: Eysenck's theory of the criminal personality; cognitive explanations; level of moral reasoning and cognitive distortions, including hostile attribution bias and minimalisation; differential association theory; psychodynamic explanations.

- Dealing with offending behaviour: the aims of custodial sentencing and the psychological effects of custodial sentencing. Recidivism. Behaviour modification in custody. Anger management and restorative justice programmes.

Offender Profiling: Top-down approach

Key terms questions

Q1 Explain what is meant by 'offender profiling'. [3 marks]

Offender profiling is the tool that is employed by police forces when they investigate serious crimes. This can take either a top-down or bottom-up approach, but both methods will examine elements such as witness reports and crime scene evidence that may have been left behind. The intention of both methods is also the same in that they aim to generate hypotheses about the probable characteristics of the offender such as their age and background to narrow the scope investigation and ultimately aid in catching the perpetrator.

Q2 Explain what is meant by 'organised type of offender'. [3 marks]

An organised type of offender refers to one of the typologies within the top-down approach to profiling and a specific modus operando that criminals may have. An organised offender usually plans their crimes and targets a specific demographic in society. In whatever offence they commit, they tend to demonstrate high levels of precision. Such elements of offending are associated with a suspect that may have an above-average intelligence, who may be in a skilled job, and who may have a family.

Q3 Explain what is meant by a 'disorganised type of offender'. [3 marks]

A disorganised type of offender refers to one of the typologies within the top-down approach to profiling and a specific modus operando that criminals may have. A disorganised offender can be considered the antithesis to an organised one as their crimes are often spontaneous and less directly targeted. In whatever offences they commit, they are not cautious about leaving any traces – for example, if a disorganised offender was to commit homicide, they may leave the body at the crime scene. Such elements of offending are associated with a suspect that may have a below-average intelligence and who may be unemployed.

Q4 Distinguish between an organised and disorganised type of offender. [4 marks]

The top-down approach classifies the modus operandi of offenders into two distinct profiles: the organised and disorganised typologies. An organised offender usually targets a specific demographic in society, and in whatever offence they commit, they tend to demonstrate high levels of precision. These offenders typically have an above-average intelligence, a skilled job, and potentially a family. A disorganised offender can be considered the antithesis to an organised one as their crimes are often spontaneous, usually leaving behind a plethora of evidence. Also contrary to the organised typology, the suspect that may have a below-average intelligence, be unemployed, and have a history of failed relationships.

Short response questions

Q5 Briefly explain how the top-down approach is used to create an offender profile. **[6 marks]**

In the top-down approach, also known as the typology approach, offender profilers who use this method will match what is known about the crime and the offender to one of two pre-existing typologies that were created from the work of the FBI's Behavioural Science Unit. This stage is referred to as 'crime scene classification' and outcomes would differ depending on what typology the offender is classified as. So, for example, if the crime scene shows few pieces of evidence left behind and shows signs of being a planned event, then the profiler can classify the perpetrator as an organised offender. This classification would inform the subsequent police investigation as it would allow police forces to narrow down their field of inquiry to someone who matches organised offender traits, such as above-average intelligence, employment in a skilled job, and the ability to maintain relationships. On the other hand, if the crime scene is riddled with evidence and appears to be spontaneous, then the profiler would classify the perpetrator as a disorganised offender. In this case, as disorganised offenders are predicted to be individuals who may have a below average intelligence who struggle to maintain employment and relationships, police forces would be able to focus their search by looking for someone who has these traits.

Q6 Explain one strength of the top-down approach to offender profiling. **[3 marks]**

A strength of the top-down approach to profiling is that there is evidence to support the organised typology. An example of this is the research of David Canter et al., who used smallest space analysis to analyse data from 100 murders in the USA. The details of each case were examined with reference to 39 characteristics thought to be typical of organised and disorganised typologies, and the findings suggested evidence that there is indeed a distinct modus operandi which fits a subgroup of offenders, allowing them to be classified within the organised typology. As there is evidence to support aspects of the top-down approach to profiling, the method benefits from having a strong degree of credibility.

Q7 Explain two limitations of the top-down approach to offender profiling. **[6 marks]**

An issue with the top-down approach to profiling is that it is only suited to crime scenes that reveal important details about the suspect, such as rape and murder. More common offences such as burglary do not lend themselves to the top-down approach to profiling because the resulting crime scene reveals little about the offender. As a result, the stage of crime scene classification, where the criminal under investigation is compared to created typologies, is not possible, and so a profile cannot be made. The approach is therefore hindered with its limited degree of practicality.

Another issue with the top-down approach is that it is based on outdated models of personality. The method itself is built on the assumption that offenders have patterns of behaviour in relation to their typology that remain consistent across all situations, but critics have argued that such an approach is unrealistic and oversimplistic. Allison et al., for example, claimed that the top-down approach sees offending as being driven by stable dispositional traits which overlooks external factors that may constantly change and impact offenders' behaviours and personality. Such critics claim that as crucial variables are undermined, the final profile may often be limited in accuracy. Ultimately, to police forces who strive to discover the true identity of a perpetrator, the top-down approach would therefore be limited in its credibility.

Essay questions

Q8 Discuss the top-down approach to offender profiling. **[16 marks]**

The top-down approach is one of two approaches that can be employed when it comes to the process of offender profiling, and is predominantly used in the United States. The approach came to existence as a result of work conducted by the FBI's Behavioural Science Unit, who created two distinct typologies through the interviews with 36 sexually motivated serial killers. These were called the 'organised' and 'disorganised' typologies and were constructed on the premise that different offenders have distinct modus operandi. These typologies form the basis of this method of profiling, hence the alternative name of the 'typology approach'.

The top-down approach consists of a number of distinct stages. The first of these is called data assimilation, and involves the gathering of crime scene evidence. The second is known as crime scene classification, where offenders are matched against the pre-existing typologies. If the crime scene lacks significant evidence, this suggests a high level of control, which would be consistent with the organised typology. Alternatively, if the crime scene is riddled with evidence, this suggests little control on the part of the offender, which would be consistent with the disorganised typology. If the offender is classified as organised, the profiler can whittle down the pool of suspects to someone who may be highly educated and in a skilled job, as these are the traits which have been associated with a typical organised offender. If the offender is classified as disorganised, the profiler can narrow down the field of inquiry to someone who may be unemployed, or has a history of failed relationships, as these are the traits which have been associated with a typical disorganised offender. The stage of profile generation would then follow, which includes the creation of hypotheses in relation to the possible perpetrator, aiding police forces to catch the offender in question.

While such an approach to profiling may seem ideal, it is actually flawed in the sense that it cannot be used for all crimes under investigation. This typology approach is only suited to crime scenes that reveal important details about the suspect – rape, for example – and is less helpful as a tool in more common offences such as burglary. This is due to the fact that typical burglary scenes do not contain many clues about the behaviour of the offender. This means that the stage of crime scene classification cannot take place and so, a profile cannot be made. Although the value of top-down profiling must not be undermined in more extreme cases where it can be applied, the fact that it cannot be used in all instances is still a limitation, especially in consideration of the fact that the bottom-up approach is applicable to a wider range of offences. Ultimately, the approach is therefore hindered by its limited degree of practicality.

In addition to its flaws in practicality, the top-down approach has another fundamental issue in that it is based on outdated models of personality. The method itself is built on the assumption that offenders have standard patterns of behaviour in relation to their typology and that these patterns remain consistent across situations, but critics have argued that such an approach is unrealistic. Allison et al., for example, claimed that the top-down approach is based on an outdated model of personality that sees offending as being driven by stable dispositional traits, which overlooks external factors that may constantly change and influence offenders. Such critics claim that as crucial variables are undermined, the final profile may often be limited in accuracy. This limits its effectiveness in aiding police forces to discover the true identity of a perpetrator, and the top-down approach would therefore be limited in its credibility.

Despite such issues, it also must be considered that some aspects of the top-down approach have actually been supported by research evidence. For example, David Canter et al., who analysed 100 murder cases in the USA and discovered that there was indeed a distinct modus operandi that was consistent to the FBI's organised typology. Although his analysis did not provide any proof for a disorganised typology, Canter's research has still confirmed the veracity of at least some aspects of the top-down approach. In any case, this ultimately contributes towards an increased degree of credibility associated with this methodology.

Offender Profiling: Bottom-up approach

Key terms questions

Q9 Explain what is meant by 'investigative psychology'. [3 marks]

Investigative psychology refers to a technique employed in the bottom-up approach, which attempts to analyse evidence through the use of statistical procedures as well as psychological theory. The main aim is to establish behavioural patterns that can be applied to many different crime scenes. There are three aspects: interpersonal coherence, which refers to the way an offender behaves at the crime scene; time and place, which can provide information about employment and location of the offender's base; and forensic awareness, where certain offenders are aware of forensic techniques and will 'cover their tracks.'

Q10 Explain what is meant by 'geographical profiling'. [3 marks]

Geographical profiling refers to a technique employed in the bottom-up approach which makes use of the location of linked crime scenes and infers the likely location of the offender's base. This is used in conjunction with psychological theory to create hypotheses about the offender's modus operandi. Canter's circle theory makes use of geographical profiling and proposed two models of offenders: the marauder, an offender who operates in close proximity to their base, and the commuter, an offender who is likely to travel a long distance. The crux of this theory is that if there are a large number of offences, the location of these offences are likely to form a circle around their usual residence.

Short response questions

Q11 Briefly explain how the bottom-up approach is used to create an offender profile. [6 marks]

The bottom-up approach makes use of investigative psychology to create offender profiles. It attempts to analyse evidence through the use of statistical procedures as well as psychological theory. The main aim is to establish behavioural patterns that can be applied to many different crime scenes. There are three aspects: interpersonal coherence, which refers to the way an offender behaves at the crime scene; time and place, which can provide information about employment and location of the offender's base; and forensic awareness, where certain offenders are aware of forensic techniques and will 'cover their tracks'.

Another aspect within the bottom-up approach to profiling is the practice of 'geographical profiling'. Through a process known as 'crime mapping', investigators use information about linked crime scenes to make inferences about the likely home of the offender. Relying on the assumption that offenders often operate close to their base, it is said that an understanding of the spatial pattern of their behaviour provides investigators with a centre of gravity, which is likely to include the offender's home in the middle of the spatial pattern. Modern takes on geographical profiling are based on Canter's circle theory, which takes into account 'the commuter', an offender who travels far from their home base, in addition to 'the marauder', who operates close to their base, just as is predicted in original models. Crucially, this aspect of bottom-up profiling focuses on identifying the whereabouts of the perpetrator, rather than their behavioural traits, which other forms of investigative psychology may do.

Q12 Briefly explain how investigative psychology is used to create an offender profile. [4 marks]

Investigative psychology involves making judgements about the personality traits or other psychological features of offenders in order to develop a profile. One way it achieves this is through the creation of a statistical database. Statistics are used to establish a baseline of behaviours likely to occur in specific crimes, including reference to common traits of offenders. Later offences can then be compared against this baseline in order to reveal specific aspects of the perpetrator, which can include details about their personality, background, or family. Another aspect within investigative psychology is considering the significance of the time and place of the offence. For example, if the time that the offence occurred was within normal working hours, a profiler may be able to discern the employment status of the offender, and considering the place of the crime may reveal details of the location of the offender's base through geographical profiling.

Q13 Outline research into geographical profiling. [6 marks]

Geographical profiling is one aspect within bottom-up profiling, with the central aim of deducing the location of the offender in question. The technique was first proposed by Kim Rossmo in 1997, and through a process known as crime mapping, it utilises information relating to the location of linked crime scenes to make inferences about the likely home or operational base of an offender. Understanding the spatial patterns of the offender's behaviour provides investigators with a centre of gravity, which is likely to include the offender's base in the middle of the spatial pattern. This original take on geographical profiling was also beneficial in allowing the investigators to predict the location of the offender's next criminal act, given how an understanding of their spatial behaviour allows for the creation of a jeopardy surface.

Modern takes on geographical profiling have been adapted to also incorporate knowledge from Canter's circle theory. Canter proposed two models of offending behaviour. The first of these is the marauder, who operates in close proximity to their home base. The second of these is the commuter, who will travel distances away from their home before they commit their offences. Ultimately, such research into geographical profiling has been significant in helping police forces locate the perpetrators they may be searching for.

Q14 Explain one strength of the bottom-up approach to offender profiling. [3 marks]

A strength of the bottom-up approach to offender profiling is that there is evidence to support specific aspects of it. For example, Samantha Lundrigan and David Canter's research strongly backs the technique of geographical profiling. By analysing 120 previous murder cases involving serial killers in the USA, the researchers discovered that most cases did in fact suggest a centre of gravity, with the offender's base located in the middle of the special pattern. As research validates the techniques employed in bottom-up profiling, the approach ultimately benefits from a high degree of credibility.

Q15 Explain two limitations of the bottom-up approach to offender profiling. [6 marks]

A limitation of the bottom-up approach to profiling is that there is evidence to challenge its effectiveness. Gary Copson surveyed 48 police forces and discovered that information provided by profilers using the bottom-up approach only led to a successful identification of the offender in 3% of cases. This suggests that the technique is very limited in its effectiveness, which ultimately hinders the credibility of bottom-up profiling.

Another major limitation of bottom up profiling – and to the specific technique of geographical profiling – is that it is limited in its practicality. Geographical profiling is only possible when there have been multiple crimes committed by the same offender, as this is the only way to develop an adequate centre of gravity which would then in turn lead to insights of the perpetrators whereabouts. This element of geographical profiling is therefore not able to determine the location of one-time offenders, ultimately limiting the use cases for bottom-up profiling.

Essay questions

Q16 Discuss the bottom-up approach to offender profiling. [16 marks]

The bottom-up method is one of two approaches when it comes to offender profiling. It is typically employed in the UK, and differs from its top-down counterpart in that it aims to develop a profile of the suspect as the investigation progresses. There are numerous techniques within the bottom-up approach, and one of the most fundamental is the process of investigative psychology. This practice allows for the development of a profile in numerous ways. An example is the use of statistical databases. Within this technique, profilers first establish a baseline of offending behaviours that are likely to co-exist across various crime scenes, including reference to the traits of these offenders. Later offences would be compared against this baseline in order to reveal specific aspects of the perpetrator in question, which can include details about their personality, background, or family.

This aspect of the bottom-up approach is strengthened by research support. David Canter and Rupert Heritage conducted a content analysis of 66 sexual assault cases, in which the data was examined using the 'smallest space analysis' technique. The process involves using a computer program to identify correlations across patterns of behaviour. This study found five common patterns of behaviour which were likely to appear in different patterns in different individual perpetrators. This is significant, as it shows how the use of statistical baselines of behaviours assist in the creation of a profile for any later offences, as it suggests that future perpetrators will also demonstrate these behaviours. As this is exactly the methodology that is employed in investigative psychology, Canter and Heritage's research confirms the bottom-up approach's veracity and ultimately increases its credibility.

Another aspect within the bottom-up approach to profiling is the practice of geographical profiling. Through a process known as crime mapping, investigators use information about linked crime scenes to make inferences about the likely home of the offender. Relying on the assumption that offenders often operate close to their base, understanding the spatial pattern of their behaviour provides investigators with a centre of gravity, which is likely to include the offender's home in the middle of the spatial pattern. Modern takes on geographical profiling are based on Canter's circle theory, which takes into account 'the commuter', an offender who travels far from their home base, in addition to 'the marauder', who operates close to their base, just as is predicted in original models. Crucially, this aspect of bottom-up profiling provides additional information concerning the whereabouts of the perpetrator rather than solely concerning their behavioural traits, which investigative psychology may do.

This aspect of the bottom-up approach is also strengthened by research support, such as that provided by Samantha Lundrigan and David Canter. The researchers collated information from 120 murder cases involving serial killers in the USA, and analysed it with the smallest space analysis technique. It was found that there was in fact spatial consistency in the behaviour of the killers, where the location of each body disposal site tended to be in a different direction from the previous sites. It was also discovered that these locations created a centre of gravity, with the offenders base located in the middle, which supports the principles of geographical profiling. Research therefore illustrates that bottom-up profiling is both accurate and effective, strengthening its credibility.

Furthermore, the bottom-up approach is valued more highly than the top-down approach due to its greater scientific basis. The argument portrayed by Canter is that bottom-up profiling is more objective and scientific, grounded in evidence and psychological theory and less focused on speculation and the use of fixed typologies. Furthermore, with the rapid advancement of AI, investigators are able to manipulate data to provide further insights that can assist in the investigation.

Despite such support for the bottom-up approach, one must also consider the evidence that challenges it. For example, Gary Copson surveyed 48 police forces and discovered that information provided by profilers using the bottom-up approach only led to a successful identification of the offender in 3% of cases. Although there is a plethora of support for aspects within bottom-up profiling, the technique is nonetheless limited in its successes. Therefore, it is hindered in its possible practical value and cannot be adopted uncritically.

Biological explanations of offending behaviour: a historical approach

Key terms questions

Q17 In relation to the Biological explanations of offending behaviour, explain what is meant by 'atavistic form'. **[3 marks]**

Atavistic form is a biological approach to offending that states that criminality and offending behaviour are the result of a given individual's evolutionary failures. According to the explanation, offenders are a primitive subspecies that are savage and untamed, meaning that they simply cannot conform to the rules imposed by modern society. This inability to assimilate oneself into a civilised society is what leads to offending behaviour.

Short response questions

Q18 Outline the atavistic form as a biological explanation of offending behaviour. **[6 marks]**

Cesare Lombroso outlined his beliefs about the causes of offending behaviour in biological terms. He stated that criminals are actually a primitive subspecies, fundamentally different from non-criminals, and he made this claim with reference to the process of evolution. Lombroso claimed that criminals lacked the same evolutionary development that non-criminals would have experienced, leading the former to naturally be more 'savage' than the latter. With a greater degree of savagery, criminals are said to experience greater difficulty in conforming to the modern rules of a civilised society, thus leading to offending behaviour.

Given the theory's belief in evolutionary differences between criminals and non-criminals, offenders are said to have distinctive physiological markers that differentiate them from normal humans. For example, murders are said to have bloodshot eyes, curly hair, and long ears, while sexual offenders are said to have glinting eyes, swollen or fleshy lips, and projecting ears. The theory of an atavistic form was the earliest and is one of the most prominent biological theories in psychologists' attempts to explain offending behaviour.

Q19 Give one strength of the atavistic form as a biological explanation of offending behaviour. **[3 marks]**

A strength of the atavistic form theory lies in the immense contributions the notion has had in the field of criminology. For instance, Lombroso's explanations shifted the paradigm of causes of offending away from theories of morality towards examining more biological explanations. This is significant because an explanation that revolves around matters such as genetics and evolution is inherently more scientific than an explanation that revolves around untestable psychological matters. Ultimately, the atavistic form theory carries value through its impacts on the field of criminology, establishing it as a field with scientific credibility.

Q20 Give one weakness of the atavistic form as a biological explanation of offending behaviour. **[3 marks]**

A weakness of the atavistic form theory is that the evidence for the concept of atavistic characteristics is often contradictory. Goring wanted to establish whether there were physical or mental differences between criminals and non-criminals, as Lombroso initially theorised. Goring conducted a comparison between 3,000 criminals and 3,000 non-criminals and found that there was no evidence that offenders are a distinctive group with differential features. However, he did suggest that many people who were criminals had a lower-than-average intelligence, which may provide some limited support for Lombroso's argument, but it does dispute the key point that criminals have different appearances.

Q21 Outline one study in which the atavistic form was investigated. Include details of what the psychologist(s) did and what they found. **[4 marks]**.

Lombroso himself conducted a study in order to assess the validity of his own atavistic form theory. In the procedure, he investigated the skulls of 383 dead criminals and 3,839 living ones, making note of those with atavistic characteristics. Through a thorough analysis, he had discovered that 40% of criminal acts are committed by those who possess atavistic characteristics, and this led Lombroso to conclude that the atavistic form is indeed associated with criminality.

Essay questions

Discuss the historical explanation of offending behaviour. **[16 marks]**

The historical explanation was that offending behaviours were caused by biological factors. This theory was proposed by Cesare Lombroso in 1876. Although it is considered outdated and contentious in the modern day, the theory at the time offered some profound insights. The theory of an atavistic form stipulates that criminals are actually a primitive subspecies that are fundamentally different from non-criminals, and Lombroso made this claim with reference to the process of evolution. He stated that criminals lacked the same evolutionary development that non-criminals would have experienced, leading the former to naturally be more 'savage' than the latter. With a greater degree of savagery, he theorised that criminals experience greater difficulty when attempting to conform to the modern rules of a civilised society, which ultimately leads to offending behaviour.

While the crux of the theory lies in offering an explanation centred around evolution, Lombroso went on to provide an account of the physiological characteristics that offenders may possess.

Given the evolutionary differences that are said to exist between criminals and non-criminals, it is indeed conceivable to claim that offenders can have distinctive physiological markers that differentiate them from normal humans. Lombroso stated that murders have bloodshot eyes, curly hair, and long ears, while sexual offenders are said to have glinting eyes, swollen or fleshy lips, and projecting ears. This specific aspect of the historical explanation was significant in laying down the foundations for the typology approach used in America, and was ultimately both a fundamental part and a rigorously tested aspect of the atavistic form theory. Lombroso tested his theory by carefully examining the facial and cranial features of Italian convicts. There were 383 dead convicts and 3,839 living, and he concluded that 40% of criminal acts were committed by individuals who had atavistic characteristics.

This historical explanation of offending behaviour is strengthened by the significant contributions it has had in the field of criminology. The theory of an atavistic form had many positive ripple effects on the pool of knowledge that contemporary criminologists had; to cite just one such example, it caused a paradigm shift. Prior to the theory, the dominant explanation of criminal behaviour centred around morality and claimed that offenders were, in essence, weak-minded humans. Such explanations carry little scientific credence and are therefore less compatible with criminology as a legitimate, scientifically backed discipline. Lombroso's explanations however, moved the focus to more biological elements. This is significant because an explanation that revolves around matters such as genetics and evolution is inherently more scientific than an explanation that revolves around untestable and unverifiable psychological matters. Ultimately, the atavistic form theory has value in its impacts on the field of criminology through the introduction of more scientific credibility.

The range of tests that have been conducted in relation to the atavistic form, however, actually work more to disprove this historical explanation. One study which goes towards refuting its specific premise was conducted by Charles Goring in 1913. He compared 3,000 criminals and 3,000 non-criminals, and reached the conclusion that there was actually no evidence that criminals are a subspecies of humans with distinctive physiological traits, thereby challenging Lombroso's ideas. Although it must be considered that there are some earlier studies that support the historical explanation – such as that conducted by Lombroso himself – later ones such as this once conducted by Goring may still carry more credence, given the more scientific nature of psychology at the time of his study. As such, the spate of reliable evidence against the atavistic form theory ultimately hinders its validity. Furthermore, the lack of a control group in Lombroso's research further weakens the research support for the atavistic approach. Unlike in Goring's research, Lombroso didn't compare his criminal sample group to a non-criminal sample, so it is possible that the significant differences he observed may no longer be present.

In addition, this same aspect of the historical explanation further suffers from the fact that it endorses a form of scientific racism. This is because many of the atavistic characteristics put forward by Lombroso – such as dark skin and curly hair – target people of African descent, which by extension implies that Africans are more likely and can be assumed to be criminals. This is why the theory can be considered contentious, and why the theory has been criticised by many other psychologists such as Matt DeLisi, who viewed Lombroso's explanation of criminality as highly discriminatory. Although the value the theory may have had to contemporary criminologists must be weighed up, it is also important to consider that the attitudes within society and within both criminology and psychology as fields of study were different then to how they are today. Such a matter is still a significant issue because the controversial nature of this biological explanation means that it simply cannot be accepted in the present-day field of the subject. Ultimately, this hinders the value that the theory has.

Biological explanations of offending behaviour: genetics and neural explanations

Short response questions

Q23 Outline the neural explanations of offending behaviour. **[6 marks]**

There are two dominant neural explanations in this biological approach to offending behaviour. The first of these centres around the role of the prefrontal cortex. It is said that reduced activity in the prefrontal cortex – an area of the brain responsible for regulating emotional behaviour – can lead to traits associated with antisocial personality disorder (APD). These may include a lack of remorse for one's actions, which is something that can encourage offending behaviour. Raine conducted many studies of the APD brain and found an 11% decrease in volume of grey matter in the prefrontal cortex in APD individuals than in controls.

The second of these explanations revolves around the role of mirror neurons – a construct in the brain responsible for controlling empathy reactions. This theory suggests that some can only feel empathy if they actually try to, and so in a natural state will lack the feeling. This phenomenon is also associated with individuals with APD, and as this lack of empathy would prevent such people from resonating with their potential victims, it could also encourage offending behaviour.

Q24 Explain one strength of the neural explanation of offending behaviour. **[3 marks]**

A strength of the neural explanation of offending behaviour lies in the range of evidence that serves to support it. For example, Adrian Raine conducted reviews which found that offenders with APD tended to demonstrate reduced activity in their prefrontal cortexes, supporting this strand of the theory. Regarding another key aspect of the neural explanation, Christian Keysers et al. illustrated how those with APD only feel empathy when they were asked to emphasise with someone, supporting the aspect which revolves around mirror neurons. Ultimately, with such evidence, one can ensure that neural explanations carry a significant degree of validity.

Q25 Explain two weaknesses of the neural explanation of offending behaviour. **[6 marks]**

One weakness of the neural explanation of offending behaviour is that it evokes a sense of biological reductionism. This is because it attributes offending behaviour solely to the role of the prefrontal cortex or mirror neurons while completely disregarding other matters which may encourage criminality. For example, the explanations make no reference to social factors, which other theories such as Sutherland's theory of differential association identify as the most fundamental factor in leading to criminality. As such, neural explanations may not offer a comprehensive view in relation to all the factors that may cause offending behaviour in the real world, making it less applicable and potentially too narrow.

Another weakness lies in the fact that neural explanations demonstrate biological determinism. Both strands of the theory state that without a fully functioning part of the brain – be this the prefrontal cortex or mirror neurons – offending behaviour is inevitable, and this is undesirable in a social sense. It means that criminals cannot take responsibility for their actions in a court of law if they are found to have such defects, and this is discordant with the belief in free will tied in with our legal system. Ultimately, this also limits the explanation's applicability to the real world.

Q26 Describe one study that has investigated neural explanations of offending behaviour.
[4 marks]

One study into the neural explanation was conducted by Christian Keyers, whose work focused on the role of mirror neurons. He showed criminals with antisocial personality disorder a film with a person experiencing pain while their brains were hooked up to a scanning device. Some were just shown the film without any prior instruction, while others were told to emphasise with the person beforehand. It was found that those in the latter group demonstrated activity of their mirror neurons and so felt empathy, unlike those in the former groups. Keyers et al. concluded that criminals with APD do indeed feel empathy if they actively try to, but they lack the capacity in its more permanent state, as is seen in ordinary people. Without having the ability to empathise as their default state, this can make them both more willing and more able to engage in behaviour – including criminal behaviour – which could hurt others.

Q27 Outline the genetic explanations of offending behaviour.
[6 marks]

The genetic explanation for offending behaviour takes an interactionist approach and is built upon the diathesis-stress model. The diathesis/vulnerability factor for criminal behaviour lies in a number of candidate genes. These include the MAOA gene – which controls dopamine and serotonin, both of which can be linked to aggressive behaviour – and the CDH13 gene – which is linked to ADHD. The theory suggests that if carriers of either gene are exposed to a significant amount of stress in whichever form – for example, by being raised in a dysfunctional family environment – then the effects of the respective genes would be realised.

Those in possession of the MAOA gene would experience abnormalities in their dopamine and serotonin system, leading to aggressive and potentially violent offending behaviour while those in possession of the CGH13 gene would become more erratic with the development of ADHD, potentially also leading to offending behaviour. Both genetic explanations offer a biological explanation for the causes of crime in similar fashion. Furthermore, it has also been found that the closer in genetic relatedness two individuals are, the more likely they are to demonstrate the same behaviours. Lange carried out a criminal twin study on 13 monozygotic (MZ) twins and 17 dizygotic (DZ) twins, where at least one of the twins served some form of prison time. He found that in 10 of the MZ twin pairs, both twins had served some prison time, while this figure was only 2 for DZ twins.

Q28 Explain one strength of the genetic explanation of offending behaviour. **[3 marks]**

A strength of the genetic explanation of offending behaviour is that there are twin studies which support the theory's claims in relation to the role of genes. Johannes Lange conducted one such study by investigating 13 monozygotic (MZ) twins and 17 dizygotic (DZ) twins. At least one of each twin pairing had served time in prison and were therefore considered criminal. Lange found that 10 of the MZ twin pairs were both offenders, while only 2 of the DZ twins could both be classified as an offender. As the MZ twins are closer genetically than the DZ twins, this demonstrates that genetic factors likely do have an influence of offending behaviour. This evidence strengthens the validity of the theory.

Q29 Explain two weaknesses of the genetic explanation of offending behaviour. **[6 marks]**

One weakness of the genetic explanation of offending behaviour is that it demonstrates biological reductionism. This is because it attributes offending behaviour solely to genetics while completely disregarding other matters which may encourage criminality. For example, the explanations make no reference to cognitive factors which some research – for instance, Kohlberg's theory of levels morality – identifies as the most fundamental factor in leading to criminality. As such, genetic explanations may not offer a comprehensive view in relation to all the factors that may cause offending behaviour in the real world, making it less practically applicable when it comes to fully explaining criminality.

Another weakness with genetic explanations lies in their sense of biological determinism. By stating that the causes of crime can only be attributed to genes associated with criminality, the theory is problematic in a social sense as it isn't compatible the current criminal justice system. An example of this can be seen in the case of Stephen Mobley. Mobley killed a pizza shop manager in 1991 and claimed he was 'born to kill' due to the presence of a 'criminal gene' as shown by his family history of violence. However, this argument was rejected, and he was sentenced to death. This suggests that in practice such a determinist position isn't favourable as it allows individuals to excuse their behaviour and escape accountability. The genetic explanations may therefore not have a practical value in the real world, in which the concept of free will and responsibility for one's actions is commonly accepted.

Q30 Describe one study that has investigated genetic explanations of offending behaviour. **[4 marks]**

Johannes Lange conducted a twin study in order to investigate genetic explanations of offending behaviour. The sample involved 13 monozygotic (MZ) twins and 17 dizygotic (DZ), where at least one twin in each pairing had served time in prison, and by definition were considered criminal. Through some simple analysis, Lange had discovered that 10 of the MZ twin pairings out of the 13 could both be classified as a criminal. Conversely, she discovered that only 2 of the 17 DZ twin pairs could both be classified as a criminal. Lange came to the conclusion that genetic factors do in fact play a role in offending behaviour.

Application questions

Long response questions

Q31 Discuss neural explanations of offending behaviour. **[16 marks]**.

There are two dominant neural explanations in this biological approach to offending behaviour. The first of these focuses on the role of the prefrontal cortex. Reduced activity in the prefrontal cortex – an area of the brain responsible for regulating emotional behaviour – can lead to varied impacts. For example, it may lead one to demonstrate a lack of empathy, or a lack of emotional responses more generally, both of which are traits associated with antisocial personality disorder (APD). This condition in itself had been found to have a link to delinquency, as individuals who feel no remorse for their actions have a greater propensity to commit crime. This correlation offers one neural explanation for offending behaviour.

The second of these explanations revolves around the role of mirror neurons – a construct in the brain responsible for controlling empathy reactions. The notion suggests that some can only feel empathy if they actually try to, and so in a natural state, they lack the feeling. This phenomenon is also associated with individuals with APD, and as this lack of empathy would prevent such people from resonating with their potential victims, it could also encourage offending behaviour.

Both aspects of the neural explanations of offending behaviour are strengthened by the range of evidence that serves to support them. For example, Adrian Raine conducted reviews which found that offenders with APD tended to demonstrate reduced activity in their prefrontal cortexes, confirming the veracity of the belief that defects in this region of the brain is associated with criminality. Christian Keysers et al.'s study supports the second strand of this explanation, as it illustrated how those with APD could only feel empathy when they were asked to emphasise with someone, supporting the aspect which revolves around mirror neurons. Although in relation to the first study, there may have been numerous other causes for APD and associated behaviours other than the prefrontal cortex, the combination of these and other research studies still demonstrate that neural explanations carry a significant degree of validity.

On the contrary, the neural explanation of offending behaviour is hindered by the fact that it evokes a sense of biological reductionism. This is because it attributes offending behaviour solely to the role of the prefrontal cortex or mirror neurons while completely disregarding other matters which may encourage criminality. For example, the explanations make no reference to social or cognitive factors, which are identified by Sutherland's theory of differential association and Kohlberg's theory of levels morality as fundamental factors in leading to criminality. As such, neural explanations may not offer a comprehensive view of all the factors that may cause offending behaviour in the real world, making it limited in how applicable it may be.

In addition, both neural explanations are further hindered by their strong sense of biological reductionism. By attributing crime to a deficient part of the brain – be this the prefrontal cortex, or mirror neurons – the theory arguably makes excuses for criminality and treats it as inevitable. It means that criminals cannot take responsibility for their actions in a court of law if they are convicted, as they can blame their actions on the role of such deficits and claim diminished responsibility. This is discordant with the belief in free will tied in with our legal system, making neural explanations inadequate and insufficient.

Q32 Discuss genetic explanations of offending behaviour. **[16 marks]**

The genetic explanation for offending behaviour postulates that potential offenders possess certain genes which predispose them to become criminals. Taking an interactionist approach with its foundations on the diathesis stress model, the theory states that the effects of these 'candidate genes' would only be realised if carriers of said genes are exposed to environmental stress. For example, one of the candidate genes the theory makes reference to is the MAOA gene, which controls dopamine and serotonin, both of which can be linked to aggressive behaviour. But a carrier of MAOA would only experience defects in their dopamine and serotonin system and would only demonstrate heightened aggressiveness (which is linked to offending behaviour) if they are exposed to stress such as that posed by a dysfunctional family, for example. In similar fashion, carriers of the CDH13 candidate gene, which is linked to ADHD, would only become more erratic with the development of the ADHD symptoms linked to offending behaviour if exposed to a significant stressor.

This genetic explanation of offending behaviour is strengthened by the fact that there are twin studies serving to support it. Johannes Lange conducted one such study by investigating 13 monozygotic (MZ) and 17 dizygotic (DZ) twin pairings in which at least one of each twin pairing had served time in prison, and were therefore considered criminal. Lange found that 10 of the MZ twin pairs were both able to be classified as offenders, while only 2 of the DZ twin pairs were both offenders. This demonstrates that genetic factors do indeed have an influence of offending behaviour. Although the fundamental issues surrounding twin studies must be considered in that it may have been the effect of the shared environment (nurture factors) that tend more prevalent amongst MZ twins, rather than the effect of shared genes (nature factors), such studies still go some way in strengthening the validity of this theory.

Despite the presence of supporting research, the genetic explanations are still flawed as a result of their biological reductionism. The theory attributes offending behaviour solely to genetics while completely disregarding other matters which may encourage criminality. For example, such explanations make no reference to cognitive factors, which Kohlberg's theory of levels of morality claims is the most fundamental factor in leading to criminality. They make no reference to social factors either, which Edwin Sutherland had claimed as most significant in his differential association theory. All of this calls into question the applicability of the theory in that, if genetic explanations do not offer a comprehensive view including all the factors that may cause offending behaviour, can the notion really be applied to the real world, or is it too simplistic? This matter must be considered, as it significantly hinders the practical value and comprehensive validity that genetic explanations hold.

In addition, genetic explanations suffer from their sense of biological determinism. By stating that the causes of crime can only be attributed to genes associated with criminality, the theory is problematic in a social sense. It allows criminals to avoid taking responsibility for their actions in a court of law if they are convicted, as they can blame their actions on the role of their genes and claim diminished responsibility. This is discordant with the belief in free will tied in with our legal system. While providing some theoretical value, it is ultimately evident that genetic explanations suffer in a practical sense. It is also reductionist in this way, as it assumes that everyone with specific genetic traits and experiences is predestined to be a criminal. This makes it impractical and incompatible with the criminal justice system in the real world.

However, there is supporting evidence for the diathesis-stress model of crime. A major study was carried out on 13,000 Danish adoptees by Mednick et al. For the purpose of the study, criminal behaviour was defined as one court conviction. When neither the biological nor adoptive parents demonstrated criminal behaviour, the percentage of adoptees that did was 13.5%. However, when either the biological or adoptive parents had convictions, this figure increased to 20% and if both had convictions, the figure was 24.5%. This may well suggest that both genetic inheritance and environmental influences play important roles in offending.

Q33 Discuss the biological explanation of offending behaviour. **[16 marks]**

There are a wide range of biological explanations that have been proposed in recent times, but the first one of these was put forward by Cesare Lombroso in 1876 in his historical explanation. His theory of an 'atavistic form' stipulates that criminals are actually a primitive subspecies that are fundamentally different from non-criminals, and Lombroso made this claim with reference to the process of evolution. He stated that criminals lacked the same evolutionary development that non-criminals would have experienced, leading the former to naturally be more 'savage' than the latter. With a greater degree of savagery, criminals are said to experience greater difficulty when attempting to conform to the modern rules of a civilised society, which ultimately leads to offending behaviour.

This particular biological explanation of offending behaviour is strengthened by the significant contributions it has had in the field of criminology. The theory of an atavistic form had many positive ripple effects on the pool of knowledge that contemporary criminologists had. Prior to the theory's development, the dominant theory to explain offending behaviour focused on morality and claimed that offenders are, in essence, weak-minded humans. Such explanations carry little scientific credence and so do not lend themselves to criminology as an empirical discipline. Lombroso's explanations, however, moved the focus from morality to biology; this is significant because an explanation that revolves around genetics and evolution is inherently more scientific from an explanation that revolves around untestable psychological matters. As the atavistic form theory was the first major biological explanation for criminality, it therefore carries immense value through its impacts in adding more scientific credibility to the field of criminology.

Later biological explanations have developed to make reference to a much wider range of biological phenomena. Some explanations say that criminality is a matter that is inherited, for example. These theories focus on the role of genes, evidently with a different nuance compared to Lombroso's initial theory from 1876. This genetic explanation for offending behaviour takes an interactionist approach with its foundations on the diathesis stress model, unlike the solely nature-based theory proposed by Lombroso. It postulates that the effects of 'candidate genes' would only be realised if carriers of said genes are exposed to environmental stress. For example, carriers of the candidate gene 'MAOA' – which controls dopamine and serotonin, both linked to aggressive behaviour – would only experience defects in their dopamine and serotonin system and the subsequent heightened aggressiveness (which is linked to offending behaviour), if they are exposed to stress such as that posed by a dysfunctional family. Such explanations illustrate how the biological explanations for offending behaviour have become more sophisticated with time.

This biological explanation of offending behaviour is strengthened by the fact that there are twin studies serving to support the importance of genetics. Johannes Lange conducted a study investigating 13 monozygotic (MZ) and 17 dizygotic (DZ) twins where at least one person in each twin pairing had served time in prison and was therefore considered criminal. Lange found that 10 of the MZ twin pairs were both offenders, while only 2 of the DZ twin pairs were both offenders. This demonstrates that genetic factors do indeed have an influence of offending behaviour. Although it is impossible to tell what is the effect of the shared environment (nurture factors) that tend more prevalent amongst MZ twins and what is the effect of shared genes (nature factors) in twin studies, such studies still go some way in strengthening the validity of this take on biological explanations.

Despite the presence of supporting research, the biological approach's genetic explanation is still flawed as a result of its biological reductionism. The theory attributes offending behaviour solely to the role of the genetics while completely disregarding other matters which may encourage criminality. For example, such explanations make no reference to cognitive factors, which Kohlberg's theory of levels of morality identifies as the most fundamental factor in leading to criminality. They make no reference to social factors either, which Edwin Sutherland claimed were most significant in his differential association theory. All of this calls into question the applicability of the theory; if genetic explanations do not offer a comprehensive view including all the factors that may cause offending behaviour, can the notion really be applied to the real world or sufficiently explain all aspects and causes of criminality? This must be considered, as it significantly hinders the value and validity that this biological explanation holds.

However, there is supporting evidence for the diathesis-stress model of crime. A major study carried out on 13,000 Danish adoptees by Mednick et al. provided evidence for this theory. For the purpose of the study, criminal behaviour was defined as one court conviction. When neither the biological nor adoptive parents demonstrated criminal behaviour, the percentage of adoptees that did was 13.5%. However, when either the biological or adoptive parents had convictions, this figure increased to 20% and if both had convictions, the figure was 24.5%. This may well suggest that both genetic inheritance and environmental influences play important roles in offending.

Psychological explanations of offending behaviour: Eysenck's theory

Key term questions

 Q34 Which two of the following statements about Eysenck's theory of the criminal personality are TRUE? **[2 marks]**

Shade **two** boxes only.

- A. The criminal personality avoids sensation-seeking situations.
- B. The criminal personality cannot be conditioned easily.
- C. The criminal personality has a high level of introversion.
- D. The criminal personality has an over-aroused nervous system.
- E. The criminal personality scores highly on neuroticism.

Short response questions

 Q35 Outline Eysenck's theory of the criminal personality. **[6 marks]**

Eysenck's theory involves two fundamental dimensions of personality: neuroticism/stability and introvert/extrovert, which are encapsulated within the notion of a criminal personality. Eysenck proposed his theory by first outlining his belief that all personalities are biological in origin, and are linked to the nervous system that one may inherit. He stated that an extraverted personality is the result of an underactive nervous system and is characterised by a constant drive to seek excitement and stimulation and a neurotic personality is characterised by nervousness and erratic behaviours. He believed that the combination of both sets of traits was associated with a criminal personality.

Eysenck also claimed that a neurotic-extravert (criminal) personality leads to offending behaviour due to the way it interacts with socialisation processes. Socialisation is the process of learning norms, values, and behaviours that occurs as children grow, including being taught how to delay gratification. Eysenck believed that the criminal personality is linked to a nervous system that makes the processes of conditioning more difficult, which would affect how well the individual could be socialised. He stated that this would lead to a greater likelihood of offending behaviour, as neurotic-extravert individuals would be more concerned with immediate gratification. As such, Eysenck's theory links offending behaviour to the personality type that one may develop, and how that personality type makes delinquent behaviour more likely.

Q36 Give one strength of Eysenck's theory of the criminal personality. **[3 marks]**

A strength of Eysenck's theory is that there is research to support it. Sybil Eysenck and Hans Eysenck compared 2,070 male prisoners' scores on the EPI – a tool to measure personality – with 2,422 male controls. It was found that the 2,070 prisoners (criminals) recorded higher scores on the Neurotic and Extravert sides of the respective dimensions than the 2,422 controls (non-criminals). This suggests that personality type may in fact be linked to or even a determinant of offending behaviour, strengthening the theory's validity.

Q37 Give two weakness of Eysenck's theory of the criminal personality. **[6 marks]**

One weakness of Eysenck's theory of the criminal personality is that its central notion of a single personality type is flawed. Critics have put forward their arguments, claiming that Eysenck's model is too simplistic. For example, Terrie Moffitt demonstrated how there are actually several criminal personality types, which differ based on variables such as how long offending persists in the individual. Furthermore, it can be seen as lacking nuance when compared to modern personality theory. For example, Digman's five-factor model of personality suggests that while extraversion and neuroticism are important, so is openness, agreeableness and conscientiousness. It can be seen now that a high score for extraversion and neuroticism alone may not mean offending is inevitable and as such, Eysenck's theory may suffer from a limited degree of explanatory power, and this hinders its overall value.

Another weakness is that Eysenck's theory is culturally biased, and this has been suggested by various research projects. For example, Cart Bartol and Howard Holanchock discovered that Hispanic and African-American prisoners tended to be less extraverted than a non-criminal control group. Eysenck's theory predicates that criminals should actually be more extraverted, and as this was not the case with this particular cultural group, the theory may not be universally applicable.

Q38 Describe one study related to Eysenck's theory of criminal personality. In your answer include information on what the researcher(s) did and what they found. **[6 marks]**

Sybil Eysenck and Hans Eysenck compared 2,070 male prisoners' scores on the EPI – a tool that measures an individual personality across the various dimensions stipulated in this theory – with 2,422 male controls. Groups were subdivided by age, ranging from 16-69 years. It was found that on measures of psychoticism, extraversion, and neuroticism, the 2,070 prisoners (criminals) of all ages recorded higher scores than 2,422 controls. The researchers concluded that there is indeed a distinct criminal personality that offenders are likely to have, which validates the theory of criminal personality.

Long response questions

Q39 Discuss Eysenck's theory of the criminal personality. Refer to evidence in your answer.

[16 marks]

Eysenck's research attempted to link the likelihood of offending behaviour to the personality type that an individual may have. Eysenck stated that personality types can be represented across two distinct dimensions – neuroticism/stability and introvert/extrovert – and the sides of these dimensions that an individual may gravitate towards depends upon the nervous system that they may inherit. For example, an individual with an underactive nervous system is predicted to form an extraverted personality, characterised by constant drive to seek excitement and stimulation. At the same time, they would also develop a neurotic personality, characterised by nervousness and erratic behaviours. Eysenck claimed that the combination of the extravert and neurotic traits would form what he referred to as the 'criminal personality type', or the 'neurotic-extravert personality'.

Eysenck's theory also provides a precise explanation regarding exactly how the criminal personality causes offending behaviour by making reference to the process of socialisation. Socialisation is the process of learning norms, values, and behaviours that occurs when children grow, and it involves being taught how to delay gratification. Eysenck believed that the criminal personality and the nervous system which leads to it make the processes of conditioning more difficult, and as a result, would lead to a failure of socialisation overall. As such, individuals would therefore be more concerned with immediate gratification. They would therefore be more likely to act on their drives to seek constant excitement due to their extraversion, and their behaviours more generally would be erratic. Eysenck stated that such matters would lead to greater instances of behaviours which one may consider delinquent or criminal.

Eysenck's theory is strengthened by supporting research evidence, such as that conducted by Sybil Eysenck and Hans Eysenck. These researchers had compared 2,070 male prisoners' scores on the EPI – a tool to measure personality – with 2,422 male controls. On measures of extraversion and neuroticism, the 2,070 prisoners (criminals) recorded higher scores than 2,422 controls (non-criminals) across all age groups. This suggests that personality type, in relation to the extraversion and neuroticism, are correlated with offending behaviour and that there may indeed be a district criminal personality type that offenders are likely to have. However, when considering the value that such evidence has in supporting the theory, it must be acknowledged that this particular study only studied men. Trying to use these findings to justify the application of a distinct criminal personality in women may give rise to beta bias, as there is no way to tell if the same principles are seen across different genders from this study. Therefore, while this study does demonstrate some validity of the theory, it does not demonstrate if it is universally applicable. But ultimately – for males at least – such research indicates that the theory holds true, thus strengthening its validity to an extent.

Evaluating such research leads one to question the applicability of Eysenck's theory, and this particular issue has actually been raised by other researchers too, such as Curt Bartol and Howard Holanchock, who brought the matter of cultural bias to the fore. The researchers had studied Hispanic and African– American offenders in a maximum-security prison in New York. They divided these into six groups based on their criminal history, as well as the nature of their offence. It was found that all six groups were less extraverted than a non-criminal control group, and this does in fact suggest a major issue in the applicability of Eysenck's theory. The notion of a criminal personality stipulates that offenders should be more extraverted – and as the Hispanic and African-American offenders in this study had not shown this, it suggests that the theory of a single personality type cannot be applied to other cultural groups. This ultimately hinders the theory in revealing its culturally relative nature. When considered with the fact that there is little proof of the theory applying to females too, it could be argued that Eysenck's theory is significantly limited because it does not explain offending behaviour in universal or absolute terms.

Further to this, Eysenck's criminal type is inconsistent with more modern personality theories. For example, John Digman's five factor model of personality suggests that alongside extraversion and neuroticism, there are additional dimensions of openness, agreeableness, and conscientiousness, which Eysenck's theory fails to account for. Given how multiple combinations are available with these newer models that Eysenck's theory does not encapsulate, our current understanding of personality is more nuanced. This means that that a high extraversion and neuroticism score alone may not mean that offending behaviour is inevitable, since scores on these other dimensions will also have an impact. Ultimately, Eysenck's theory is therefore discordant with current understanding of personality, and so is limited in its application, not only in gender or cultural terms as mentioned, but also as it may be oversimplistic.

Psychological explanations of offending behaviour: Cognitive explanations

Key term questions

Q40 Explain what is meant by 'hostile attribution bias' and give one example. **[3 marks]**

Hostile attribution bias refers to the cognitive distortion that causes the misinterpretations of actions; it causes one to perceive the behaviours of others as being hostile, even if they are not in reality. This often leads to a disproportionately violent response, which many would consider offending behaviour. For example, an individual when presented with an image of an emotionally ambiguous facial expression will be likely to perceive this as angry and react accordingly.

Q41 Explain what is meant by 'cognitive distortions' and give one example. **[3 marks]**

Cognitive distortions refer to the errors that one may experience when processing information, and it is characterised by faulty thinking. One example is the distortion of minimalisation, whereby offenders downplay or deny the seriousness of their actions by downplaying the effects, rationalising their reasons as to why they committed the crime, or trivialising their actions.

Q42 Explain what is meant by 'levels of moral reasoning' and give one example. **[3 marks]**

Levels of moral reasoning is a concept proposed by Kohlberg which claims that people's decisions and their beliefs about whether actions are right or wrong can be summarised in stages of moral development. The higher the level, the more sophisticated the reasoning becomes. Level one is known as preconventional, level two is conventional, and level three is postconventional. Criminals are most likely to be classified at the preconventional level. This is characterised by a need to gain rewards and avoid punishment.

Q43 Outline one cognitive distortion shown by offenders who attempt to justify their crime. **[2 marks]**

One cognitive distortion shown by offenders attempting to justify their actions is minimalisation. This involves the faulty thought process of an offender dampening the severity of their crimes in their head, and justifying their actions as if they could be regarded as acceptable.

Short response questions

Q44 Explain how the cognitive distortion can be used to explain offending behaviour. **[6 marks]**

Cognitive distortions refer to the errors in an individual's information processing system, and two distinct types can be used to explain offending behaviour. The first of these is known as hostile attribution bias. This is the misconception that individuals may experience when perceiving the behaviours of others. They often perceive simple actions as hostile, and this commonly leads to a violent response, which may include violent offending behaviour. Schonenberg and Aiste showed 55 violent offenders images of emotionally ambiguous facial expressions and compared their responses with a matched control group. They found that the violent offenders were significantly more likely to perceive the expressions in these images as angry than the control group were.

The second known cognitive distortion in relation to offending behaviour is the notion of minimalisation. This is when individuals dampen the severity of their own actions in their minds. They wrongly think that typical criminal behaviours are acceptable in certain contexts, and this belief often leads them to demonstrate offending behaviours more commonly than those without the distortion.

Q45 Give one strength of the cognitive distortion theory as an explanation of offending behaviour. **[3 marks]**

One strength of the notion of cognitive distortions pertain to the real-life applications that they have had in recent times. When it comes to rehabilitation, cognitive behavioural therapy (CBT) tends to be the dominant approach, and with emerging research into cognitive distortions, the therapy has now incorporated new techniques. For example, in the adaptations of CBT, offenders are encouraged to establish a more realistic view of their crimes, which serves to tackle the effects of minimalisation in causing offending behaviour and aims to reduce recidivism. Evidently, the cognitive distortion explanation therefore benefits from an immense degree of practical value.

Q46 Give two weakness of the cognitive distortion theory as an explanation of offending behaviour. **[6 marks]**.

One weakness of the cognitive distortion theory when it comes to explaining offender behaviour is that it does not account for any nature– or nurture-based factors which may contribute to offending behaviour. Mednick et al.'s study of 13,000 Danish adoptees provided evidence that both nature and nurture could play a role in offending behaviour. Within their study, they found that when neither the biological nor adoptive parents demonstrated criminal behaviour, the percentage of adoptees that did was 13.5%. However, when either the biological or adoptive parents had convictions, this figure increased to 20%, and if both had convictions, the figure was 24.5%. This study demonstrates that there may be more elements leading to criminality than simply cognitive distortions, making it potentially too narrow.

Another weakness of this theory is that it does not fully explain how these cognitive distortions develop. It is therefore more descriptive than explanatory, although it could be used to explain recidivism or how some offenders may avoid taking accountability for their actions. For instance, a rapist who uses minimisation to claim that their victim wasn't really harmed or that they were 'asking for it' may be more likely to reoffend than one who accepts that their behaviour was both harmful and unacceptable. However, this only works for explaining continued offending and not what led to them committing the offence to begin with, making this an incomplete explanation.

Q47 Explain how the 'level of moral reasoning' can be used to explain offending behaviour.
[6 marks]

Lawrence Kohlberg's theory of levels of moral reasoning attempted to determine the likelihood of an individual engaging in offending behaviour. Kohlberg's model put forward three distinct levels and at each of these levels, an individual's reasoning is said to be based upon different principles. The first level is known as the preconventional level, in which an individual's reasoning is driven by a desire to avoid punishment while maximising personal gain. As such, an individual who is classified at the preconventional level would commit crimes if they can avoid consequences and reap rewards, making offending behaviour more likely. The second level is the conventional level, in which an individual's reasoning is driven by a desire to uphold the social order or to be regarded as a 'good boy/girl'. The third level is the postconventional level, in which an individual's reasoning is driven by a strong set of ethical principles. Both of these later stages would make offending behaviour less likely. Ultimately, Kohlberg's theory explains offending behaviour in terms of the individual having a low level of moral reasoning and predicts that criminal acts are less likely once an individual reaches a higher level.

Q48 Give two weaknesses of the level of moral reasoning theory as an explanation of offending behaviour.
[3 marks]

A strength of Kohlberg's level of moral reasoning theory is that there is research to support it. For example, Emma Palmer and Clive Hollin conducted a study which involved comparing moral reasoning between 210 female non-offenders, 122 male non-offenders, and 126 convicted offenders. They found that the 126 convicted offenders showed less mature and a lower level of moral reasoning than the non-offender groups, and this is consistent with Kohlberg's ideas that a lower level of moral reasoning correlates to a greater likelihood of offending behaviour. The theory therefore benefits from a strong degree of proven validity.

Q49 Give two weaknesses of the level of moral reasoning theory as an explanation of offending behaviour.
[6 marks].

Kohlberg's level of moral reasoning theory has been weakened in recent times due to the range of arguments that have emerged against it. For example, John Gibbs challenged Kohlberg's principle of a postconventional level, stating that it was culturally biased in favour of western democratic cultures and that it did not represent a natural maturational state of cognitive development. He offered an alternative approach comprising an 'immature' and 'mature' level of reasoning – which were essentially identical to the preconventional and conventional levels – but in any case, this ultimately hinders the theory in revealing how some elements of it can be considered unrealistic, or even biased towards some cultures.

Another weakness of the level of moral reasoning theory is that there is limited research support. Though Emma Palmer and Clive Hollin carried out a study which showed that the 126 offenders in their sample demonstrated a lower level of moral reasoning than the 322 non offenders, it must be considered that this proof is correlational and not causational. There may have been a range of other factors that had caused the convicts to commit their crimes, quite distinct from their moral reasoning level. Because this has not been controlled for as a variable, it cannot be definitively claimed that this theory fully explains offending behaviour.

Long response questions

Q51 Discuss one or more cognitive explanations of offending behaviour. **[16 marks]**.

There are two distinct cognitive explanations behind the phenomenon of offending behaviour, the first of which is encapsulated in Kohlberg's theory of moral reasoning. He put forward a model whereby the degree of an individual's morality can be segregated into three levels. Each level is characterised by distinctive principles upon which reasoning is based. The first of these is the preconventional level. Kohlberg stated that an individual at this level would base their reasoning on a drive to avoid punishment and a desire for personal gain – and so would be more likely to commit a crime if they could any consequences or can gain some sort of reward out of it, be it monetary or otherwise. The second level is the conventional level, and reasoning at this stage is said to be based on a desire to be seen as 'good boy/girl' or to uphold social order. The final stage put forward by Kohlberg was the postconventional level, in which an individual's reasoning revolves around a strong set of ethical principles. Kohlberg's theory states that a lower level of moral reasoning makes criminal behaviours more likely. Conversely, at the higher levels, where an individual's reasoning is said to be more sophisticated, Kohlberg's theory predicts that offending behaviour is less likely.

Kohlberg's unique cognitive explanation is strengthened by the research that has emerged in support of it. For example, Emma Palmer and Clive Hollin conducted one such study, which involved comparing the moral reasoning of 210 female non-offenders and 122 male non-offenders to 126 convicted offenders. The findings illustrated that the 126 convicted offenders were classified as having a lower level of moral reasoning than the non-offender groups, providing proof for the notion that lower moral reasoning is associated with criminal behaviours. However, it must be considered that this proof is correlational. There may have been other factors which caused the convicted offenders to commit the crimes they did, quite distinct from their levels of moral reasoning. But the research still adds an element of reliable support to Kohlberg's theory nonetheless, especially considering how the entirety of the large sample of offenders demonstrated low levels of moral reasoning – while the entirety of the non-offender samples conversely demonstrated higher levels. Given this, Kohlberg's theory of moral reasoning benefits from a significant degree of proven validity.

In spite of this support, Kohlberg's level of moral reasoning theory has still been weakened in recent times through the range of arguments that have emerged against it. For example, John

Gibbs challenged Kohlberg's principle of a postconventional level, stating that it was culturally biased in favour of western democratic cultures, and did not represent a natural maturational state of cognitive development. He offered an alternative approach comprising solely the 'immature' and 'mature' levels of reasoning – which were identical to the preconventional and conventional levels, but did not include an equivalent of the postconventional level. Gibbs' counter argument also carried great credence as it is consistent with other existing theories pertaining to morality – such as Jean Piaget's theory of moral development. As such, these alternative models must also be considered as opposed to solely accepting Kohlberg's theory.

Furthermore, the presence of individual differences and inconsistencies based on the type of offence weakens the explanatory power of cognitive explanations. Thornton and Reid found that individuals who committed crimes for financial gain – for example, robbery – were more likely to be classified as pre-conventional compared to those who commit impulsive crimes where reasoning is not apparent. Beyond this, Langdon suggested that intellect may be a better predictor of criminality than moral reasoning. This is because research has shown that people of very low intelligence had lower levels of moral reasoning, but they were also less likely to commit a crime, whilst Kohlberg would predict that they'd be more likely to be criminal as they operate at the pre-conventional level.

The other distinct cognitive explanation of offending behaviour is the concept of cognitive distortions. Two examples of faulty thought processes which link to why individuals may engage in criminal behaviours are hostile attribution bias and minimisation. Hostile attribution bias is the phenomenon whereby individuals perceive the actions of others as being hostile, even if they are actually neutral. This often leads to a disproportionately violent response, thus offering an explanation for violent offending behaviour in particular. Minimisation, meanwhile, is the way that individuals dampen the severity of their actions in their minds. For example, they may consider actions such as child molestation as less problematic than it actually is, leading the individual to commit such acts and tell themselves that it is not something that can be considered 'criminal'. This distortion offers an explanation of a wider range of offending behaviours, but recent research such as that conducted by Howard Barbaree offers some evidence that suggests that it is a stronger explanation for cases of sexual offences than explaining all offences equally.

The benefit of cognitive distortions being used to explain offending behaviour is that it has more real-life applications than other cognitive theories, such as that which was put forward by Kohlberg. For example, the predictions of the theory have been incorporated into cognitive behavioural therapy (CBT), which is the dominant approach when it comes to the rehabilitation of criminals. Particularly, it has led to a new stage in the process where offenders are encouraged to establish a more realistic view of their crimes, which serves to tackle the effects of minimisation in causing offending behaviour. This practice has been shown to correlate to a reduction in reoffending, illustrating the practical benefits and proven validity of the cognitive distortion explanation. However, this explanation is more useful in preventing repeat offenses rather than preventing or explaining crime before it occurs.

Psychological explanations of offending behaviour: Differential association theory

Key term questions

Q52 Briefly outline differential association theory as an explanation for offending behaviour.

[3 marks]

Differential association theory is a theory proposed by Edwin Sutherland, which claims that the likelihood of engaging in offending behaviour is dependent on the norms and values of an individual's social group. If a particular social group values deviant behaviour, for example by seeing drug use and dealing as a positive, it means that individuals who are part of it are more likely to engage in that behaviour.

Short response questions

Q52 Give one strength of the differential association theory as an explanation of offending behaviour.

[3 marks]

A strength of Sutherland's theory of differential association is that it has brought about a positive shift in focus in the perception of offending behaviour. In earlier times, the general paradigm was that offending behaviour was due to nurture factors, as evidenced by Lombroso's atavistic theory, which offered no realistic means to prevent criminality. Sutherland's contrasting focus on nurture factors – an individual's social environment – benefits from the fact that it offers more scope to tackle the issues of offending behaviour. For example, it suggests that behavioural therapies can tackle the impacts of acquired criminal attitudes, which provides a practical value in how we deal with offending behaviour.

Q53 Give two weakness of the differential association theory as an explanation of offending behaviour.

[6 marks].

One fundamental weakness of the differential association theory is that it is highly difficult to test. Sutherland stated that criminal behaviours can be predicted by measuring the amount of criminal acts learnt and criminal behaviours acquired. Questionnaires could be employed to assess these variables, but the issue of social desirability bias would be especially prevalent in this case, and so it is difficult to accurately measure. In this case, as it also relates to the offending behaviours of others in an individual's social circle, people may be hesitant to disclose information which could implicate those they care for. As such, credibility of differential association theory is difficult to establish.

Another weakness of the differential association theory is that there may be individual differences. While Sutherland may be correct in his understanding regarding the effects of learnt criminal acts and attitudes on increasing the likelihood of offending behaviour, it must be considered that this principle may not apply to all. For example, despite having exposure to such criminal factors, if an individual chooses not to offend by their own volition, then despite what the theory would suggest, offending behaviour simply would not occur. Ultimately, differential association is therefore limited in value due to the broad generalisations it makes, its failure to acknowledge free will, and its failure to acknowledge the importance of an idiographic approach in offending behaviours.

Q54 Describe one study related to the differential association theory as an explanation of offending behaviour. In your answer include information on what the researcher(s) did and what they found.

[6 marks]

Farrington et al. performed a longitudinal study to investigate the influence of life events and family background and identify the risk and factors predicting offending and antisocial behaviour. 411 boys from white working-class backgrounds in London were selected for study, and 365 of them were interviewed again 40 years later. Having a parent or other family member who had been convicted of a crime was identified as one of the factors in whether these boys went on to engage in offending behaviours, while those without convicted family members were less likely to engage in crime. This supports the differential association theory as it demonstrates that the norms and values of the people around you influence how likely someone is to engage in offending behaviour.

Application questions

Q55 Researchers studied three generations of several families, noting the frequency and type of offending. They found that sons and grandsons of offenders often committed similar crimes themselves. The researchers also interviewed people who knew the families, such as friends and neighbours. Most f0riends and neighbours were not concerned by the offending behaviour, and some said it was a good way to behave in the circumstances.

Describe differential association theory in the context of offending. Refer to the study above in your answer.

[8 marks]

Differential association theory was a product of Edwin Sutherland's attempt to offer a scientific explanation for offending behaviours. It states that there two factors that increase the likelihood of criminality which an individual can learn. The first of these is criminal attitudes. Sutherland stated that when an individual socialises into a group, they may well be exposed to various attitudes, broadly generalised into the 'pro-crime' and 'anti-crime' categories. The theory suggests that if the number of 'pro-crime' attitudes outweigh those of the 'anti-crime' category, then the members of that group would develop a mindset that is more gravitated towards criminal ideologies. For example, with reference to the study above, the sons and grandsons of the offenders would have been likely to demonstrate deeper criminal attitudes, considering the prevalence of those attitudes in their immediate family group.

The second learnt factor is that of criminal acts. Sutherland stated that this process can occur in various forms, ranging from imitation even to direct tuition from other offenders. Referring to the study above, the sons and grandsons within the sample would have been likely to possess a range of skills linked to criminal acts, given the amount of people they may have imitated or received direct tuition from. Ultimately, the theory states that these learnt criminal ideals, in combination with the skills to act upon these, would logically culminate into an increased likelihood of offending behaviour. This explains the findings of the study above, as the sons and grandsons often committed crimes themselves

Long response questions

Q56 Discuss the differential association theory as an explanation of offending behaviour.
[16 marks].

Edwin Sutherland put forward his theory of differential association to explain offending behaviour as part of his attempt to establish a set of scientific principles regarding criminal behaviours. The general premise is that individuals learn matters that eventually encourage offending behaviour by associating with different people. The first of these variables is criminal attitudes. According to proponents of differential association, when an individual socialises into any given group, they will be exposed to a range of different ideologies. Some of these can be classified as 'pro-criminal', while some can be considered 'anti-criminal'. The theory suggests that if the number of attitudes in the former category outweigh those in the latter, then an individual will develop and foster the values of a criminal. According to Sutherland, this is one of two prerequisites that lead to offending behaviour.

The second of these factors are criminal acts themselves. In this same group, if 'pro-criminal' attitudes are in the majority and crime is committed, an individual may well pick up various skills in relation to committing crimes. For example, they could learn from others through the process of imitation, or they could even acquire knowledge of these same acts via direct tuition from others. According to the theory of differential association, if an individual develops the ideologies of a criminal and learns the correct skills to act on these views, then there would be a greater likelihood of offending behaviour. This is how Sutherland had explained the phenomenon of criminal acts.

One strength of the differential association theory is that there is evidence to support it. Sutherland stated that the family is critical in determining whether the individual will engage in offending behaviour. If the family supports criminal activity, portraying it as legitimate and reasonable, this can have a large effect on that individual's moral system. This is supported by the fact that offending often runs in families. A study by Mednick et al. showed that boys who had criminal adoptive parents and non-criminal biological parents were more likely to engage in offending behaviour than boys whose adoptive and biological parents are non-criminal, suggesting that the nurture factors are important in determining criminality.

However, based on the premise above, the idea that offending runs in families could support alternative psychodynamic explanations rather than the psychological differential association theory. If moral behaviour is controlled by the superego, this develops through the resolution of the Oedipus or Electra complexes which results in identification with the same-sex parent. The superego becomes the internal voice holding an individual's morals and values. However, if a child develops an inadequate superego through identification with a criminal parent, they will also become a criminal as they have internalised the parent's criminal morals and values.

Despite this, differential association theory is brought down by the fact that the premise is rather difficult to test. It is hard to measure, for example, the number of pro-criminal attitudes a person has, or has been exposed to. Questionnaires could be used as a means to assess these variables, but the data from such methods would be significantly limited in validity due to the issue of social desirability bias. Individuals may lie about their attitudes or lie to protect those who they associate with. Without being able to measure these factors in an accurate way, it is difficult to know when criminal behaviours start to become more likely. In spite of its self-proclaimed scientific approach, this means that differential association may actually hold little value in allowing us to predict criminal or offending behaviour. As such, its scientific credibility is hindered.

In addition to this issue, the differential association theory is also flawed through its failure to account for individual differences. While Sutherland may be correct in his understanding regarding the effects of learnt criminal acts and attitudes on increasing the likelihood of offending behaviour, this principle may not apply to all. For example, despite having exposure to such criminal factors, if an individual chooses not to offend by their own volition, their environment cannot force this to occur. Differential association is therefore limited in value due to the broad generalisations it makes, the fact it does not take into account the role of free will, and its failure to acknowledge the importance of an idiographic approach to offending behaviours.

Psychological explanations: Psychodynamic theory

Short response questions

Q57 Outline the psychodynamic explanation of offending behaviour. **[6 marks]**

One of the most prominent psychodynamic explanations for offending behaviour revolves around the notion of an inadequate superego. Ronald Blackburn, who proposed this theory, stated that if the superego that an individual possesses is somehow deficient, offending behaviour would be near inevitable. This is because without a superego with enough influence to punish the ego with guilt for criminal behaviours, the id – which would drive an individual to engage in such acts – would be given free rein to act on its impulses.

Blackburn proposed three different types of deficient, or inadequate superegos, with each having its own unique link to offending behaviour. These included: the weak superego, which would develop if a child's same sex parent is absent during the phallic stage so the child can't internalise a fully formed superego; the deviant superego, which would develop if the child internalises the superego of a criminal parent; and the overharsh superego, which would drive an individual to offending via the unconscious influence of the superego's need to punish.

Another psychodynamic explanation of offending behaviour comes from John Bowlby. He believed that maternal deprivation leads to affectionless psychopathy, which in turn leads to criminality in adulthood. In this hypothesis, both juvenile and adult criminality is said to originate from inadequate bonds with the mother in childhood.

Q58 Give one strength of the psychodynamic explanation of offending behaviour. **[3 marks]**

A strength of the psychodynamic explanation put forward by Bowlby in relation to maternal deprivation is that there is research to support it. The 44 thieves study showed that 12 of the 14 thieves who were classified as having affectionless psychopathy – the trait that Bowlby linked directly to offending – had all experienced maternal deprivation in their early lives. This supports the theory's proposition that maternal deprivation leads to a psychological trait which then encourages criminal behaviours, validating the central tenet of the maternal deprivation hypothesis.

Q59 Give two weakness of the psychodynamic explanation of offending behaviour. **[6 marks]**.

A weakness of psychodynamic explanations – specifically those in relation to the role of an inadequate superego – is that they tend to suffer from a lack of falsifiability. Given the unconscious nature of the concepts within Freudian theories, applications to crime, such as those considered in relation to the role of the superego, are not open to empirical testing. This is an issue: because of the absence of supporting evidence, arguments concerning the role of the inadequate superego can only be judged on their face value, rather than scientific worth. In other words, there is absolutely no way to assess if the inadequate superego actually holds significance in relation to offending behaviour, and this hinders the value of psychodynamic theories. Because of this, the notion contributes little to our understanding of crime, why criminal behaviours may originate, or how we can prevent it.

Another issue with psychodynamic explanations – here in relation to maternal deprivation – is that the research proof is limited. Although there may be evidence provided by Bowlby's 44 thieves study, there are numerous matters which hinder the value of this. To cite just one example, interviews with parents were used to assess periods of separation, and given the retrospective nature of this data, there may have been many inaccuracies. If the data was flawed in such regards, then we cannot be sure if long periods of separation – i.e. maternal deprivation – actually leads to a trait that encourages criminality. Ultimately, with little proof of credibility for the maternal deprivation hypothesis, the value the theory holds in being linked to offending behaviour is significantly limited.

Q60 Outline one study related to the psychodynamic explanation of offending behaviour. In your answer include details of what the researcher(s) did and what was found. **[4 marks]**

John Bowlby conducted the 44 thieves study to assess the validity of his psychodynamic explanation linking maternal deprivation to offending behaviour. In the procedure, A sample of 88 children were selected from the clinic where Bowlby worked, and these included 44 were juvenile thieves as well as 44 other children who served as controls. Each child had their IQ tested by a psychologist who also assessed the child's emotional attitudes towards these tests. At the same time, a social worker interviewed a parent to record details of the child's early life (e.g., periods of separation). The psychologist and social worker made separate reports, and Bowlby then compared the findings of these two reports. The reports showed that 14 of the 44 young thieves displayed signs of affectionless psychopathy, with 12 of these having experienced prolonged separation from their mothers during the critical period. From this comparison, Bowlby came to the conclusion that maternal deprivation leads to the psychological trait of affectionless psychopathy, which subsequently encourages offending behaviour.

Long response questions

Q61 Discuss two or more psychodynamic explanations of offending behaviour. Refer to evidence in your answer. **[16 marks]**

The psychodynamic approach provides two main explanations for offending behaviour. The first of these was put forward by Ronald Blackburn, and revolves around his notion of an 'inadequate superego'. The premise of Blackburn's theory was that if a child internalises a superego that is deficient in any regard, then criminal behaviour would be near inevitable. This is because of the implications that a weak superego would lead to. With an inability to punish the ego with a strong enough sense of guilt, the id – which would be responsible for offending impulses – would be given free rein to act on its desires, thus leading to offending behaviour. Significantly, Blackburn's psychodynamic explanation offers insights into three distinct types of inadequate superego. The first of these is the weak superego, which is said to form if the child's same-sex parent is absent during the phallic stage of development. The second is the deviant superego, which forms if the child inherits the superego of a criminal parent: a superego with criminal ideologies. And the third of these is the overharsh superego, which differs slightly from its counterparts as it is the strong need for punishment that a strong superego entails which unconsciously drives an individual to engage in criminal behaviours.

While Blackburn's psychodynamic theory seems to offer a detailed insight into the causes of offending behaviour, his explanations still suffer from being gender biased. All Freudian theories – such as that which Blackburn based his explanation on, in relation to the structure of personality – claim that girls develop weaker superegos than boys. This understanding stems from the belief that girls do not experience castration anxiety, and so are supposedly put under less pressure to identify with their mothers, which leads to a weaker superego. On the other hand, boys – who do experience castration anxiety – are conversely assumed to be put under immense pressure to identify with their fathers, leading to a stronger superego. Given this, Blackburn's explanation leads to the logical assumption that girls would engage in offending behaviour more often that boys, considering their presumably weaker superego. It is this aspect of the theory that carries a multitude of issues. First and foremost, the extent of alpha bias here is problematic. But even more significantly, the alpha bias here lacks any validity. This can be evidenced by the fact that, in 2021, there were more male inmates in prisons across England and Wales (75,000 male inmates) relative to female inmates (3,200) – despite the theory suggesting that it should be the opposite. The theory of an inadequate superego is therefore hindered by the fact that it unjustly biases against females, and wrongly states that they are more likely to commit crime, which weakens the validity and credibility.

Another major limitation to Blackburn's psychodynamic explanation of offending behaviour, particularly for those relating to a deviant superego, is that they are highly deterministic. In essence, the theory stipulates that a child with a criminal parent would grow up to be a criminal through their inherited deviant superego, and that this will be passed down through generations. The theory thus suggests that one criminal bloodline could be trapped in their offending ways, and this is undesirable, as accordingly, there would be no other way except from means such as adoption to a 'normal' family to escape this endless cycle. And even with such a solution, it may be deduced from the theory that the absence of a biological father would still lead to the child developing a weak superego, possibly leading them to become a criminal in any case, and beginning the same endless cycle once more. As the theory suggests that offending behaviour can be hardwired into a person's fate and that nothing can be done about it, its credibility is significantly hindered. The hard determinist stance of this theory makes it less desirable than those which are less determinist and which contrastingly offer a solution to criminal behaviours – such as differential association – and have room for both free will and rehabilitation.

The second major psychodynamic theory in relation to offending behaviour was put forward by John Bowlby, and is linked to his hypothesis of maternal deprivation. According to Bowlby, if a child experiences prolonged separation during their 'critical period', then this would have a lasting psychological impact on them. One of these impacts is the development of a trait known as 'affectionless psychopathy'. Bowlby stated that the main characteristic of this trait is a lack of empathy, and his theory suggests that this in itself increases the likelihood of criminal acts.

This is because it means that they would lack remorse for their actions, and therefore may be more likely to engage in acts of delinquency – including offending behaviour. Unlike Blackburn's theory – which cannot be tested due to the unconscious nature of all the concepts it draws upon – Bowlby's psychodynamic explanation benefits from some research support. Through his 44 thieves study, which involved interviews with the thieves and the parents, Bowlby found that 14 of the 44 young thieves showed affectionless psychopathy, and of this sample, 12 had experienced prolonged separation from their mothers during the critical period. This study validates the notion that maternal deprivation can lead to affectionless psychopathy, which can in turn lead to criminal behaviours such as thieving. There are however issues with this study in that it relies upon possibly inaccurate retrospective and there is the possibility of researcher bias, given how Bowlby conducted the study to prove his own theory. These factors must be considered when weighing up its value as evidence, but the 44 thieves study still goes some way in strengthening the validity of this particular psychodynamic theory.

However, there is evidence contradictory to the idea that maternal deprivation causes offending behaviour. Lewis analysed data drawn from interviews with 500 young people and found that maternal deprivation was an extremely poor predictor of future offending. Even if there is a link between children who experience prolonged or frequent separation from their primary caregiver and offending behaviour later on in life, this link is not definitively causal. There may be many other reasons for this apparent link, for instance biological explanations or the differential association theory. An alternative could also be that maternal deprivation is only one of the reasons and there is a combination of factors that play a role in causing offending behaviour.

Dealing with offending behaviour: custodial sentencing

Key term questions

Q62 Explain what is meant by 'recidivism'. [3 marks]

Recidivism in simple terms means reoffending. More specifically, it refers to a tendency to relapse into a previous mode of behaviour. In the context of offending, this would mean a convicted offender committing a repeated offence.

Short response questions

Q63 Outline the aims of custodial sentencing. [6 marks]

There are a range of aims to custodial sentencing – the practice of sending a convicted offender to an institution based on their crime – and one of these is to create a deterrence. The act of sending a criminal to prison not only serves as a general deterrence to others in society from committing crime, but also as an individual deterrence. This is because, as behaviourists would suggest, people are less likely to repeat a behaviour if faced with a punishment.

Incapacitation is another aim, whereby offenders are taken out of society in order to prevent any risk of recidivism, with the motive of protecting the public. Retribution is another key aim of custodial sentencing, whereby sending a criminal to an institution serves to exact revenge on behalf of the public. Logically, the level of punishment invoked by retribution should be directly proportionate to the severity of their offence. The final aim of custodial sentencing lies in a need to rehabilitate convicted offenders. Some believe that spending time in prison serves to reform a criminal, as during their time there they would have the opportunity to access treatment programmes and time to reflect on their actions.

Q64 Outline two psychological effects of custodial sentencing. [6 marks]

One psychological effect of custodial sentencing is the issue of stress and depression that inmates may suffer through. The sheer stress of a prison experience has been proven to be directly linked to depression, evidenced by the higher suicide rates within such institutions in comparison to rates in wider society. This is a problem in of itself, but it also increases the risk of psychological disturbance post-release.

Institutionalisation is another psychological effect. This refers to the phenomenon whereby inmates become so accustomed to prison life, that they simply cannot function adequately when they get released. This can make them more likely to reoffend, both because of the difficulties in thriving in normal employment and because it allows them to return to the familiar environment of prison.

Q65 Outline research into recidivism. [6 marks]

Recidivism refers to a tendency to relapse into a previous mode of behaviour – and in the context of offending, this would mean a convicted offender committing a repeated offence. Recidivism is indeed a common practice, and this can be evidenced by government statistics. For example, the UK Ministry of Justice in 2013 said that 57% of UK offenders will reoffend within a year of their release. In addition, in 2007, 14 prisons in England and Wales recorded reoffending rates of over 70%, suggesting issues within the key aims of custodial sentencing in relation to deterrence and rehabilitation.

There are multiple theories for why recidivism occurs, which will depend on what people perceive as the cause of criminal behaviour. For instance, Hollin suggested that recidivism is caused by institutionalisation, where prisoners come to view prison as 'home' and cannot adapt to life outside of prison. In contrast, Coid et al., found that offenders who received treatment for their mental health while in prison were 60% less likely to reoffend than untreated offenders, which suggests that prisons could be used more effectively to reduce recidivism.

Q66 Give two weaknesses of custodial sentencing as a method of dealing with offending behaviour. [6 marks]

One weakness of custodial sentencing as a method of dealing with offending behaviour is that it may actually increase recidivism rates rather than reducing them. This is because whilst offenders may learn legitimate skills during their time in prison that may be of use when they're released, they may also undergo a more dubious education. This may be the case as incarceration with more serious offenders gives younger inmates the opportunity to learn new criminal skills. This may then counteract any rehabilitation attempts, making recidivism much more likely.

Another weakness of custodial sentencing is that, due to individual differences, we are unable to definitely say whether it will have an effect on recidivism rates. Whilst prison may be psychologically challenging for many people, we cannot assume all will share the same experience. Different prisons will have different regimes, and experiences will also be dependent on the length of someone's sentence and their previous experience. Therefore, it is difficult to make conclusions about whether custodial sentencing is able to prevent offenders as a whole from reoffending.

Application questions

Q67 Following a series of riots in cities all over England, a politician was interviewed on the radio. He said, 'Rioters and looters should be sent to prison. We must send a clear message that this sort of behaviour is not acceptable. Society expects such behaviour to be severely punished.'

Briefly discuss two roles of custodial sentencing identified in the politician's statement.

[4 marks]

One fundamental role of custodial sentencing is to act as a deterrent to other potential criminals. The sheer notion of being sent to prison, for example, creates a general deterrence, as 'a clear message that crime is not acceptable' and will be punished, as the politician states. This makes potential offenders in a given society more unwilling to engage in offending behaviour in order to avoid any possible punishments.

Another role of custodial sentencing is retribution. As the political correct states, 'society expects such behaviour to be severely punished', and the act of sending a criminal to jail serves to exact this sort of revenge on behalf of society. This is given the fact that convicted offenders suffer for their acts via long periods of time spent in jail and the removal of the privileges associated with freedom.

Another politician also took part in the radio interview. She argued, 'The people were rioting for a reason. They were angry with the police and lost control.'

Q68 Outline and briefly discuss one treatment programme for people who offend because they are angry.

[4 marks]

One treatment programme that is apt for people who offend because they are angry is anger management. This is a form of cognitive behavioural therapy (CBT) which involves three distinct stages. The first is cognitive preparation, where offenders are instructed to think back and realise exactly which scenarios trigger aggressive thoughts and responses. These could include heated conversations with police officers, for example. The stage of skill acquisition follows this, and involves offenders being taught techniques to avoid such aggressive responses. These may be behavioural techniques involving better conversational skills, which may prove applicable to the aforementioned scenario in relation to the people's anger felt towards police officers. The final stage is that of application practice, where they put the skills learnt in the second stage to practice in a mock situation which would typically reflect one outlined in the first stage.

Successful application of skills would lead to a reward in order to positively reinforce what has been learnt, leading to less aggressive responses, and therefore less offending in people who might previously have engage in delinquency such as rioting because of their anger.

Long response questions

Q69 Describe and evaluate research on custodial sentencing and its effects on recidivism.

[16 marks]

Custodial sentencing – that is, the act of sending an offender to a suitable institution – has been researched in great depth. It is said that if a convicted offender is punished for their acts, then by the laws of behaviourism, that offender will experience an individual deterrence, and would therefore abstain from any later motives to engage in offending behaviour. This phenomenon occurs because the given criminal would know that any later acts they engage in could be punished, thus creating a drive to avoid such punishment. In relation to recidivism, the logical conclusion is that by providing a deterrent through punishment, a convicted offender is ultimately less likely to reoffend.

While this may be a straightforward principle, in reality, this is refuted by both research and government statistics. For example, the UK Ministry of Justice claimed in 2013 that 57% of UK offenders tended to reoffend within a year of their release. Davies and Raymond also researched this concept and found that prisons do little to rehabilitate or deter offenders, which suggests that custodial sentences do not actually reduce the chances of recidivism. With conflicting judgements between the logical conclusions from behaviourist theories on custodial sentencing, to what real-world evidence actually suggests, the claim that custodial sentences will reduce recidivism loses some of its validity.

However, this may not hold true universally. For example, while 14 prisons in England and Wales reported recidivism rates at over 70% in 2007, official statistics from Norway demonstrate that their nation has the lowest rates of reoffending in all of Europe. Given the possible differences between the UK's prison system and Norway's system, this suggests that there are some contexts in which recidivism is reduced through custodial sentencing, which supports the behaviourist theory. Differences that may exist between Western prison systems and those from other parts of the globe mean that neither governmental nor psychological research into how effective custodial sentences are at preventing recidivism can be applied to every culture universally.

Other areas of research into custodial sentencing have focused upon the effects of a criminal's time in a given institution, specifically how that may act on their psyche. Research has shown, for example, that stress and depression are common byproducts of spending time in prison, evidenced by the fact that suicide rates tend to be higher in prison population than in the general population more broadly. In relation to what this research suggests when considering recidivism, it provides one with the understanding that custodial sentencing – with these effects – actually increases the risk of reoffending. This is ultimately due to the fact that an unstable mental state has been positively correlated with offending behaviour. This is particularly true if one considers the diathesis-stress model, which states that exposure to high levels of stress can activate genetic traits associated with particular types of offending behaviour.

Research into the psychological effects of custodial sentences has uncovered objective evidence linking . The Prison Reform Trust in 2014 for example, had shown that amongst women offenders who have gone through the process of custodial sentencing and have ended up in prions, 25% of these suffered psychosis. Amongst the male samples, the figure stood at 15%. For both men and women, these figures are notably higher than in the general population. This validates the notion that custodial sentencing can create an unstable mind, which is in itself a factor leading to an increase risk of recidivism. This is supported by research from Coid et al., who found that offenders who received treatment for their mental health while in prison were 60% less likely to reoffend than untreated offenders.

However, neither study proves causation; it is possible that those who commit offences are more likely to be mentally ill than the general population and that it may not have been the custodial sentence which caused this. If this is the case, their custodial sentence may not have had any impact on their likelihood of reoffending. In the cases of those who received mental health treatment while incarcerated, their custodial sentences seem have decreased the likelihood of them continuing to offend on their release, which suggests that custodial sentencing can – if done right – be a useful tool in preventing recidivism.

Another piece of research worthy of consideration when it comes to custodial sentencing is Hollin's work around institutionalisation. He found evidence to suggest that prison became 'home' to some prisoners and that in some of these cases, the fact that they received three meals a day, a bed and companionship was more than they had prior to or would have after leaving prison. In these cases, recidivism is more likely as prison is seen as the better option. This suggests that custodial sentences only work to deter recidivism if the conditions an individual would experience outside of prison are significantly preferable to what they'd experience inside.

Q70 Discuss the psychological effects of custodial sentencing. [16 marks]

Recent research has revealed a broad variety of psychological effects that can result from custodial sentencing. One of the major effects is stress and depression. It is said that the long periods of time a convicted offender may be in prison can induce significant amounts of stress on that person, particularly given how systematic and strict prison life can be. This in itself is the primary effect, but the crux of the issue lies in how this often leads to depression experienced by the convict. And this effect is as common as it is extreme, which is proven by the significantly higher rates of suicide within prisons when compared to rates within in general population. This particular psychological effect can even be linked to greater instances of recidivism, given how an unstable mind may lead to erratic and potentially criminal actions.

This particular psychological effect has been supported by studies on numerous accounts. The most prominent of these was Zimbardo's prison experiment, which demonstrated that even under an artificial prison situation, an inmate will experience immense stress. This is evident in the fact that the study had to be ended prematurely as a result of the psychological harm that was experienced by the sample. However, it must be considered that the Stanford Prison Experiment only involved a small sample of 24 participants. In addition, all of these participants were males living in America, which may mean that the findings cannot be applied universally to women or people from other cultures. Ultimately, while the universality of the research is not directly evidenced, the study still serves to validate the theory that prison conditions lead to prisoners experiencing extreme stress.

A study conducted by the Prison Reform Trust in 2004 provides more proof for this effect, and it provides evidence for women as well as men. Through basic analysis, the Prison Reform Trust had released figures showing how 25% of women and 15% of men in prison experienced symptoms that pertained to forms of psychosis, including depression, both at higher rates than the general population. This study indicates that both males and females are subject to adverse psychological effects from custodial sentencing. However, it does not prove causation; it may be that individuals who are experiencing or who are already predisposed to mental illness (as suggested by the diathesis-stress model) are more likely to engage in offending behaviours. For this reason, it can only definitively prove a correlation between psychological effects and incarceration, not that the latter causes the former.

In addition to the prominent psychological effect of stress and depression, there are a number of other effects that may accompany this. These include institutionalisation: when a convicted offender becomes so accustomed to the norms of prison life that they are no longer able to function adequately when released, and prisonisation: the phenomenon whereby prisoners are socialised into adopting an inmate code. This code may include understanding of specific behaviours that could be rewarded inside prison – such as smuggling cigarettes – and the praise this may earn a convict from other inmates. If this inmate code is internalised by the prisoner, they may demonstrate the same behaviours on the outside when they are released with the belief that they can be rewarded, even if they are considered undesirable within mainstream society.

Although the aforementioned psychological effects of custodial sentencing may indeed exist, it must be considered that the concept is flawed in not recognising the potential individual differences. For example, different prisons in different nations can have different prison regimes, which could in turn result in different psychological effects. For example, a stricter prison in one part of the world may induce more stress than a prison in another end of the globe, which may allow for greater freedoms, and thus less stress. As such, the concept of psychological effects of institutionalisation is by no means universal, and cannot be said to apply equally to all custodial sentences. This means that the value is therefore limited, as the research is typically only into the psychological effects specific regimes rather than custodial sentences as a whole.

Dealing with offending behaviour: behaviour modification

Key term questions

 Q71 Explain what is meant by behaviour modification in custody. Use examples in your answer.
[3 marks]

Behaviour modification in custody is a practice employed in order to deal with offending behaviour. Based on the premises of operant conditioning, it works by negatively reinforcing undesirable behaviours – for example, by punishing infringement of rules within a prison – and conversely positively reinforcing desirable behaviours – for example, by rewarding acts which abide by prison rules through a token economy system.

Shor term response

 Q72 Describe behaviour modification in custody.
[6 marks]

Behaviour modification in custody as a means of dealing with offenders who have been imprisoned is a practice that is built upon the behaviourist approach. In essence, it aims to positively reinforce desirable behaviours, and negatively reinforce undesirable ones with the hope that those amongst the latter are continually demonstrated, and those amongst the latter stop occurring. The most prominent system that operates on this notion of behaviour modification is that of token economies. Prisons that adopt a token economy system positively reinforce what the institution views as good behaviour by inmates by providing tokens that can later be exchanged for privileges. They negatively reinforce what the institution views as bad behaviour by inmates given how the lack of tokens – and, by extension, the lack of the privileges that can be purchased with them – is considered a punishment. The tokens are seen as secondary reinforcers as they only gain value once the prisoner understands it can be exchanged for a reward (primary reinforcer).

Q73 Give one strength of the use of behaviour modification as a means of dealing with offender behaviour.
[3 marks]

A major strength of behaviour modification as a means of dealing with offending behaviour lies in the ease with which it can be employed. Methods such as token economy systems require no level of expertise, which other means may require, meaning that any prison worker can enforce the system. This means that behaviour modification is low cost, as specialists do not need to be employed. This makes behaviour modification a very practical and valuable approach to dealing with offender behaviour, as the financial and training requirements to employ it are very low, yet it has still shown to be effective.

Q74 Give two weaknesses of the use of behaviour modification as a means of dealing with offender behaviour. **[6 marks]**

One significant weakness of using behaviour modification such as token economies to deal with offender behaviour is that they can be considered unethical. This is given the fact that, firstly, it is a compulsory system. Inmates within a prison lack the free will to reject the system, because if they do, they would only suffer from a possible lack of privileges. Secondly, critics such as Moya and Achtenberg have drawn upon the dehumanising nature of the entire premise. Inmates are essentially treated as subjects of a Skinner's Box or like dogs to be trained, and putting individuals down to this level raised moral questions. Ultimately, this unethical nature of behaviour modification hinders the credibility that this means of dealing with offenders carry.

Another weakness is that behaviour modification methods hold little rehabilitative value, as Ronald Blackburn stated. This is because while positively reinforced acts may improve an inmate's behaviour within the institution, there are no guarantees that this learned behaviour will continue once they are released from prison and no longer receive the rewards. For example, a reinforced drive to abide by prison rules will not necessarily reduce the risk of recidivism, as following laws in the outside world is not positively reinforced. Ultimately, behaviour modification is therefore hindered in its practical value, as it is not feasible to continue to positively reinforce behaviours once a prisoner is released.

Q75 Describe one study that has investigated behaviour modification in custody. Refer to what the researcher(s) did and what they found. **[6 marks]**

One study conducted by Tom Hobbs and Micheal Holt investigated the effectiveness of behaviour modification in custody. They had gathered a sample of young delinquents to serve as the participants for this study – and had divided these individuals across four behavioural units. The first three were the experimental groups in this investigation and were subject to token economies, while the fourth group served as the control group and therefore not subjected to token economies. Over a period of time, the researchers observed a significant difference in positive behaviours amongst the experimental group relative to the control group. This led Hobbs and Holt to conclude that behaviour modification is indeed an effective way to deal with offenders in custody.

Long response questions

Q76 Describe and evaluate the use of behaviour modification in custody as a means of dealing with offending behaviour. **[16 marks]**

After the processes of custodial sentencing, within the institution that a given convict may be sent to, officers must employ means of dealing with the potential offending behaviour that individuals may demonstrate while in custody. Behaviour modification is one of these means, and has risen to particular prominence in prisons. This approach of dealing with offending is built upon the behaviourist principles of learning and works by reinforcing desirable behaviours, while punishing those which may be considered undesirable. The main way prison officers do this by implementing a 'token economy' system.

Within this token economy system, desirable behaviours such as following prison rules are rewarded with a physical token, which is considered the secondary reinforcer. These tokens can subsequently be used by inmates to gain access to a range of privileges such as extra food, extra time on the phones, or similar rewards. These privileges are the primary reinforcer. Through this process, the given behaviour is positively reinforced, creating an incentive for the prisoners to repeatedly engage in whatever actions are considered desirable by the institution in the knowledge that it will be rewarded. Conversely, undesirable behaviours would be reprimanded by a deduction in tokens, and this would negatively reinforce the given action as the loss of privileges can be considered a punishment, which discourages the behaviour. With this system, inmates can be kept in line and any potential offending behaviour could be kept to a minimum through their desire to be rewarded and avoid punishments.

The approach of dealing with offending behaviour through behaviour modification is strengthened by the sheer ease with which it can be employed. Given the nature of the system, no particular level of expertise is required for token economies to run as they should. Any typical prison guard could manage the giving and taking of tokens based on the desirable or undesirable behaviour demonstrated by inmates without specialist knowledge. In tandem to this, the approach costs very little to implement, as there is no need for expensive equipment, rigorous training, or to pay specialists. The ease of using and implementing the system – in both actual and economic sense – ultimately gives behaviour modification as a means of dealing with offender behaviour greater practical value.

On the contrary, some such as Ronald Blackburn argue that behaviour modification as a means of dealing with offending behaviours is not practical when it comes to rehabilitation. He argued that while the reinforcement of actions that abide by prison laws may well improve an inmate's behaviour within the institution, this does not guarantee that their behaviour would improve once they are released. This is particularly true as it is not feasible to continue to reward these desirable behaviours in the same way outside of a prison environment, and the lack of reward may mean that offenders have less motivation to continue to behave as they did when they were rewarded for their actions.

Additionally, targeted behaviours for reinforcement may not link to specific offending behaviours, meaning that the risk of recidivism could be high. Even if token economy systems in some institutions can be sophisticated enough to target and negatively reinforce matters such as violent offending, there's still no guarantee that the set of learnt actions would last in the long run, post-release. Indeed, if offending is actually positively reinforced in the outside world – which it often is with matters such as elevated status, money, etc. – the effects of the token economy system may well be reversed, as the rewarded behaviour becomes the undesirable one. Although it must be considered that the method is practical in its ease of use, the fact that behaviour modification is impractical in guaranteeing desirable outcomes after a prisoner's release carries much more weight, so ultimately, it only provides short-term rather than long-term benefits and cannot be the sole method of dealing with offender behaviour.

Further to this, behaviour modification is also fundamentally flawed in an ethical sense. This is given the fact that, firstly, it is a compulsory system. Inmates within a prison lack the free will to reject the system, because if they do, they would only suffer from a possible lack of privileges. Secondly, critics such as Moya and Achtenberg have correctly drawn upon the dehumanising nature of the entire premise. Inmates are essentially treated as subjects of a Skinner's Box or like dogs to be trained, with the reinforcing and punishing of various behaviours, and one can claim that putting individuals down to this level is problematic from a moral standpoint. This could make behaviour modification an ideologically inappropriate method of dealing with offender behaviour, regardless of whether it provides any behavioural changes or not.

However, in relation to the argument of no ability to reject the system, one must consider the question of whether free will should be a privilege that is open to prisoners. Indeed, retribution is one of the central aims of custodial sentencing, and allowing an inmate to reject token economy systems can be considered contrary to the notion of exacting revenge upon them by making them suffer. In similar fashion, dehumanising prisoners could be another means of achieving this aim of retribution. With both of these matters in mind, it can be argued that behaviour modification is not an unethical method of dealing with offending behaviour. It may just be a method that abides strongly with what custodial sentencing truly stands for. But the case put forward by those arguing in favour of the position of inmates may be more significant, especially when acknowledging that this supposedly unethical method may not even rehabilitate the prisoner for the long-term. As such, regardless of the contentious nature of this debate, it could still be argued that the ends still do not justify the means.

Dealing with offending behaviour: Anger management

Key term questions

Q77 Identify two methods of dealing with offending behaviour. **[2 marks]**

Behaviour modification, and anger management.

Short response questions

Q78 Outline anger management as a treatment for offenders. **[6 marks]**

Anger management as a treatment for offenders is built on the belief that anger triggers violent offending behaviour. As such, its central aim is to teach offenders techniques to control this anger which, by extension, would prevent recidivism. This form of cognitive behavioural therapy (CBT) works in three stages. The first of these is cognitive preparation, in which the offender is made aware of situations that trigger anger and that their anger responses are irrational. The second stage is skill acquisition, which involves offenders being taught a range of techniques to deal with the anger-inducing situations that were established in the first stage. These methods involve cognitive techniques that draw upon positive self-talk, behavioural techniques that focus on more effective communication systems, and physiological techniques that draw on meditation. Regardless of the techniques that are taught, all aim to challenge the irrational anger. The concluding stage is application practice, in which a mock scenario is created, often mimicking one that was the focal point of stage one. The intention is that offenders apply the techniques to deal with certain situations that they learnt in stage two in order to cement them into the offenders usual behaviour. Successful application of the skills and a lack of anger in response to the mock scenario will be rewarded with the hopes of positively reinforcing this outcome so that they will act the same way in real-life scenarios.

Q79 Give one strength of anger management as a treatment for offenders. **[3 marks]**

A strength of anger management as a treatment for offenders is that it acknowledges the significance of levels of explanation through its methodology. It draws upon ideas in the cognitive level in the first stage of cognitive preparation, it makes reference to the behavioural level with the skills that may be acquired in stage two, and it utilises the social level in the role play that forms a part of stage three. Given this multidisciplinary nature, one can claim that anger management is one of the few treatments for offenders that recognise multiple complex factors that may constitute criminal activity. This holistic take in treating offending ultimately gives anger management a greater degree of practical value in that it may be more applicable to the real world as a result of its recognition of multiple the real-world factors that lead to crime, as opposed to focusing solely on one cause.

Q80 Give two weaknesses of anger management as a treatment for offenders. [6 marks]

One weakness of anger management as a treatment for offenders is that its effectiveness is limited in the long-term. This claim was made by Ronald Blackburn – who acknowledged the methods effectiveness in the short-term – but pointed to the lack of evidence for long-term effectiveness. He stated that, given the artificial nature of the role plays within stage three, the therapy fails to reflect all the triggers that may be present in a real-life situation. This would mean that any reinforced skills to deal with triggers of anger may be superficial and hold little value in helping the individual to cope with real anger-inducing situations, thus still leading to potential offending. This argument calls into question the rehabilitative value that anger management holds as a treatment for offending, undermining its main goal.

Another weakness of anger management as a treatment for offenders is anger may not always be the cause of offending. Loza and Loza-Fanous carried out a range of psychometric tests on samples of offenders classified as violent and offenders classified as non-violent and found no differences in the levels of anger between these two groups. Furthermore, the concept of organised offenders further weakens the idea of anger the motivator for offending, as these individuals maintain a high level of control. It is primarily impulsive crimes involving disorganised offenders which are committed out of anger and therefore not all offenders will require anger management, which suggests that it is only appropriate for some offenders and that other treatments are necessary to treat other types of offenders.

Q81 Describe one study that has investigated anger management as a treatment for offenders. [6 marks]

Julia Keen et al. conducted a study with the aim of assessing the effectiveness of anger management as a treatment for offenders. The researchers followed a sample of young offenders between the age of 17 and 21 who took part in a nationally recognised anger management programme known as the National Anger Management Package. This methodology of treatment included eight two-hour sessions of the standard three stage procedure, where the first seven sessions were held over a span of three weeks, with the last being held a month later. The findings demonstrated that by the end of the time period, offenders reported increased awareness of their own difficulties in managing anger, and that they had widened their capacity to control this anger. The researchers concluded that anger management treatment can indeed reduce anger, which can in theory, reduce the likelihood of violent offending behaviour.

Q82 Explain the difference between anger management and behaviour modification in custody as ways of dealing with offending behaviour. **[4 marks]**

Anger management is a rehabilitative form of treatment that aims to tackle the issue that many view as the direct cause of violent offending behaviour: the emotion of anger. It takes more of a cognitive approach to dealing with offending, given its use of cognitive behavioural therapy (CBT) – and its stages teaching participants to recognise and deal with irrational anger responses. Behaviour modification, on the other hand, is a means of managing the behaviour of prisons in custody, and unlike anger management, the method offers little rehabilitative value in actually reducing offending behaviour. Also in stark contrast, behaviour modification is centred more on the behavioural approach through its use of token economies and the system within it of positive and negative reinforcement. While anger management provides skills to allow an individual to manage the emotions that drive their behaviour, behaviour modification provides rewards for desirable behaviour and punishments for undesirable behaviour.

Long response questions

 Q83 Discuss the use of anger management as a treatment for offenders. Refer to evidence in your answer. **[16 marks]**

The treatment of anger management is built on the premise that irrational anger reactions lead to violent offending behaviour. As such, the methodology of these techniques takes the form of a cognitive behavioural therapy (CBT), which seeks to replace this irrational response with one that is considered more rational. This is done in three fundamental stages.

First is the stage of cognitive preparation, which requires the offender to recall past experiences and their patterns of anger within these. Here, they are taught by the therapist to identify specific situations which induce anger and are informed of the fact that these aggressive responses are irrational in most cases.

This is followed by the second stage of skill acquisition, in which the offender is taught a range of techniques to deal with the particular situations discovered in stage one, with the aim of instilling more rational responses. These could include cognitive techniques which focus on positive self-talk, physiological techniques that centre on meditation, or behavioural techniques which focus on improving communication.

The stage of application practice concludes this intricate process, and as the name suggests, it involves putting the skills acquired in stage two into actual practice. Importantly however, these skills are not tested in real-life scenarios, but artificial ones created by the therapist, often constructed to mimic the scenarios that were the product of stage one. When the offender correctly uses the acquired skills, they are rewarded in some form in order to positively reinforce these responses, with the hope that this would cement the more rational set of responses to replace aggressiveness. The goal is that this would prevent violent offending behaviour in the long-term and encourage them to use these techniques in real-life scenarios.

Anger management as a treatment for offending is strengthened by the presence of supporting research. Jane Ireland for example, conducted a study comparing the progress of two groups of offenders – one which took part in an anger management programme (the experimental group), and another who received no treatment whatsoever (the control group). After the experimental group had completed twelve sessions, outcomes were assessed using an interview, questionnaire and a behavioural checklist by prison officers observing the sample. It was found that 92% of the experimental group showed an improvement, compared to 0% of the control group. This suggests that anger management is indeed an effective means of treatment.

However, when considering the value of this study as supporting evidence, many factors must be considered. First of all, the self-report techniques that were used may well be inaccurate due to social desirability bias, and this would hinder the credibility of Ireland's study. Secondly, it is only logical that a group who received some sort of therapy would show more improvements than a group who received no therapy, so it cannot be said with conviction that anger management is a more effective means of treatment than alternatives; simply that it is better than nothing. Ultimately, research does suggest that anger management is a valid treatment for offenders, but it must be acknowledged that this is only to a limited degree.

Despite the arguably valid nature of the method, anger management is still less than ideal due to its significant costs. The typical programme requires the involvement of highly trained specialists in dealing with violent offenders, which can be extremely costly to prisons which often have limited funding. This is problematic, as it means that anger management simply cannot be employed in every institution. This provides a stronger case to employ alternative means such as behaviour modification, which are more cost-effective and therefore more accessible.

While it has issues in terms of practicality, anger management as a means of dealing with offending is strengthened in a theoretical sense by its more comprehensive nature. It recognises the importance of many levels of explanation in its treatment, including cognitive, behavioural, and social elements. In real life, all of these levels contribute to offending behaviour in one way or another, and the fact that the treatment addresses them in its method suggests that it is more holistic than others. This makes it more applicable to the real world in comparison to more reductionist approaches such as behaviour management, which increases the value of anger management as a means of dealing with offending.

However, it may also be the case that anger doesn't always cause offending. Loza and Loza-Fanous carried out a range of psychometric tests on samples of offenders classified as violent and offenders classified as non-violent and found no differences in the levels of anger between these two groups. Furthermore, the concept of organised offenders suggests that irrational, explosive anger is not always the cause of offending, as these individuals maintain a high level of control. It is primarily impulsive crimes involving disorganised offenders which are committed out of anger, and therefore not all offenders will require anger management, which suggests that it is only appropriate for some offenders and that other treatments are necessary to treat other types of offenders.

Dealing with offending behaviour: Restorative justice

Key term questions

 Explain what is meant by 'restorative justice'. **[3 marks]**

Restorative justice is a system for dealing with criminal behaviour that particular focus on the rehabilitation of offenders. The process involves the offender reconciling with their victims with the intention that a combination of awareness of their crime and the greater empowerment of the victim leads the criminal to change their ways. There is a supervised meeting between the two with a trained meditator, which allows the victim the opportunity to express the emotional distress caused by the event and for the offender to see the consequences of their actions.

Short response questions

 Outline what is involved in 'restorative justice' programmes. **[6 marks]**

The restorative justice programme first involves the shift of emphasis from the idea that a given offender has committed a crime against the state (by breaking laws) to the idea that the offender has committed a crime against an individual victim. With this principle underpinning the system, restorative justice programmes facilitate supervised meetings between the offender and 'survivor', (the preferred term for victim within many of these programmes). The aim of this is for conversation to lead the offender towards accepting responsibility for their actions and feeling the guilt associated with this, given how they would bear witness to the survivor's distress. It is hoped that such epiphanies would rehabilitate the offender and encourage them to change their ways.

Importantly, restorative justice programmes have a number of variations in relation to the process. Others involve the transfer of money from the offender to the survivor to reflect that damage that may have been done by the former. Some involve the offender actually fixing any damage (to property) that they may have caused. Crucially, these replacements for the face-to-face interactions that are entailed in the standard version all still involve the focus on acceptance of responsibility and aim to foster positive change for the offender.

Q86 Give one strength of 'restorative justice' programme as a treatment for offenders. **[3 marks]**

One strength of restorative justice programmes is that they are very flexible. Given the available variations, particular forms of restorative justice can be administered in order to ensure that the programme is as tailored as it could be to the scenario or the offender in question. All of these variations can also be employed in different settings: for example, they can take place in schools as well as in cases where adult offenders have broken laws. All of this adds significant practical value to the restorative justice programme as a treatment as for offenders, given how versatile it is.

Q87 Give two weaknesses of 'restorative justice' programme as a treatment for offenders.
[6 marks]

A major weakness of the restorative justice programme is that it relies upon the intentions of the offender. The scheme may work if the offender goes into the programme with feelings of remorse and a drive to improve their ways, but it would be problematic if they register for the scheme with ulterior motives in mind. They may have no genuine feelings of regret, and may look merely to exploit the system, for instance as a convenient means to avoid prison time. If such motives are at the fore of the offender's mind, the scheme would have no rehabilitative value in the long run, and this significantly hinders the credibility of restorative justice programmes.

Another weakness of restorative justice programmes lie in the scorn they tend to receive from society. Despite the proven evidence that the scheme does reduce recidivism, the general public still views restorative justice as insufficiently punishing the offender. This is an issue, as, it means that restorative justice programmes fail to meet one of the central aims of custodial sentencing: retribution. Such matters ultimately hinder the credibility and the public support for restorative justice programmes as a treatment for offenders, which could also mean that victims themselves do not wish to take part in the process.

Q88 Describe one study that has investigated restorative justice as a way of dealing with offending behaviour.
[6 marks]

Sherman and Strang conducted a content analysis of 20 studies in the USA, Australia and the UK comparing re-offending rates of those who had undergone restorative justice versus offenders who hadn't. It was found that re-offending rates were significantly lower for those who had engaged in restorative justice than the control group. It was also found that restorative justice was more effective when the victim was an individual rather than multiple people or a business, and was especially effective when involving violent or property crimes such as burglary. However, they also discovered that it was not effective in all cases, and was more effective in providing relief to emotional distress of the victim rather than preventing reoffending.

Application questions

Q89 Jack has been convicted of a burglary and attends Betterway centre for young offenders. At Betterway, case workers aim to change how offenders think about their crimes by involving victims. The case workers encourage offenders to consider the wider effects of their crimes and appreciate how they should make up for what they have done wrong.

Which way of dealing with offending involves victims? **[1 mark]**

Restorative justice.

Q90 Referring to your answer to Question 31, describe what a case worker might recommend in order for Jack to deal with his offending. **[4 marks]**

The case worker may recommend Jack to partake in one of Betterway's restorative justice programmes to deal with the consequences of his burglary offenses. The programme – if it follows a typical format – would involve Jack meeting one of his victims, in this case, someone who he may have stolen from in the past. The case worker could hold supervised conversations which would lead Jack to 'consider the wider effects of his crime' in relation to the emotional impacts on the victim, for example. The hope is that this would lead Jack to take responsibility for his actions and alter how he views his crimes, ultimately reforming him. Alternatively, the programme may involve some sort of financial restitution with Jack making payments to the victim to make up for what he has done wrong and to compensate for the stolen items.

Long response questions

Q91 Describe and evaluate restorative justice programmes. **[16 marks]**

Restorative justice programmes are a means of treating offending behaviour which have become more prominent in recent times. The basic principles shift the emphasis away from the notion that an offender has wronged the state (by breaking the law) to a belief that the offender has harmed an individual. As such, the focus of restorative justice programmes differs from other methods of dealing with offenders. Rather than seeking to imprison the offender for breaking the law, such schemes aim to encourage the individual to bear witness to the consequences of their actions. This is done by fostering active interaction between the criminal and the 'survivor' (the term used for 'victim' in these programmes) with the hopes that this would ultimately reform them.

The actual process of restorative justice programmes differ depending on the case, but the typical format involves a number of basic steps. A supervised meeting between the offender and survivor is held where the survivor is given the chance to express the hardships caused by the offender. The combination of witnessing the anguish of the survivor and the possible acceptance of responsibility that these exchanges are said to bring about should, in theory, rehabilitate the criminal. If successful, the process would foster a sense of positive change for the offender and would allow them to become a law-abiding member of society, treating offending behaviour for the longer term.

One of the strengths of restorative justice as a method of dealing with offenders is the research support it has, particularly around decreasing recidivism. Sherman and Strang's content analysis of 20 studies across the USA, the UK and Australia found that offending rates were significantly lower when offenders had engaged in restorative justice programmes compared to those who had not. Similarly, Latimer et al. found that restorative justice programmes led to lower rates of recidivism than traditional methods of dealing with offenders, such as custodial sentencing and probation. This study also found that restorative justice led to greater satisfaction from both the victim and the offender than these other methods.

However, the scheme is weakened by the fact that such outcomes strongly depend upon the motives of the criminal. If the offender registers for the scheme with the intention of reforming themselves and a genuine drive to make amends for their past actions, then restorative justice programmes will reduce the possibility of re-offending for that given individual. However, if the offender registers for the scheme with an ulterior motive – for example, using it as a means to avoid prison – and has no genuine desire to take responsibility for their actions, restorative justice programmes will not facilitate any sense of real rehabilitation. The latter is particularly possible in instances where the criminal suffers from affectionless psychopathy. In such cases, restorative justice as a means of dealing with offending and preventing recidivism is uncertain, and there is no real way to determine which participants are engaging in good faith. This is a potential limitation of the progammes, as it is almost impossible to know for certain who would be a good candidate.

Another issue with restorative justice is public attitudes towards such programmes. Prisons in most parts of the world, and capital punishment in other regions that still employ it – are widely supported due to the belief that the degree of suffering it inflicts upon the offenders is desirable in making criminals pay for their actions. Given the much smaller degree of suffering it inflicts upon the offenders, restorative justice and its central aim of rehabilitation can be seen as insufficiently punishing and therefore failing to fulfil the role of retribution or deterrent. This ultimately hinders the credibility of restorative justice programmes as a means of dealing with offending and may make people reluctant to engage with or support them. Because they require participation from the victims of crime, this then limits their practicality if many victims are or could be unwilling to engage in the process.

However, a benefit of restorative justice programmes run is that they are very flexible. Given the amount of variations, particular forms of restorative justice can be administered in order to ensure that the programme is as tailored as it could be to the given scenario and the offender in question. A thief may be made to provide financial compensation to in addition to having conversations with the survivor, to further demonstrate that they accept responsibility. A juvenile offender can experience the programme at a school instead of a prison, as it works in any setting. This flexibility means that it is not a one-size-fits-all approach, and therefore can be used for multiple types of offences rather than, for example, just those motivated by irrational anger (as is the case for anger management programmes).

Answers to identification questions

Psychological explanations of offending behaviour: Eysenck's theory

Q34   B, E

AQA Psychology
Brilliant Model Answers

Are you aiming for the BEST grades possible? If you have answered yes, then look no further... these books are just what you need!

Written by experienced teachers and examiners, our series of books will allow you to learn, revise and organise your knowledge and understanding.

- Knowledge is precise and concise
- A well-structured and student friendly layout and organisation.
- Thorough and developed evaluation (the bit everyone needs extra help with).

Psychologyzone series
psychologyzone.co.uk

Sociologyzone series
sociologyzone.co.uk

ISBN 978-1-906468-10-1

9 781906 468101

www.ingramcontent.com/pod-product-compliance
Lightning Source LLC
Chambersburg PA
CBHW080239040426
42333CB00045BA/2467